IMAGINING A NEW NATURAL HISTORY

Imagining a New Natural History

Latin American Cultural Production in the Anthropocene

Edited by Nicolás Campisi
and Lucas Mertehikian

UNIVERSITY OF FLORIDA PRESS
Gainesville

This book will be made open access within three years of publication thanks to Path to Open, a program developed in partnership between JSTOR, the American Council of Learned Societies (ACLS), University of Michigan Press, and The University of North Carolina Press to bring about equitable access and impact for the entire scholarly community, including authors, researchers, libraries, and university presses around the world. Learn more at https://about.jstor.org/path-to-open/

Cover: Debora Hirsch. Plant (*Cypripedium dickinsonianum*), 2024. Chromogenic print, Plexiglas sandwich, float frame. 35 × 35 cm / 36.7 × 36.7 cm (framed). © Debora Hirsch.

Copyright 2026 by Nicolás Campisi and Lucas Mertehikian

All rights reserved

Published in the United States of America

31 30 29 28 27 26 6 5 4 3 2 1

DOI: https://doi.org/10.5744/9781683405559

Library of Congress Cataloging-in-Publication Data
Names: Campisi, Nicolás editor | Mertehikian, Lucas, 1986– editor
Title: Imagining a new natural history : Latin American cultural production in the Anthropocene / edited by Nicolás Campisi, Lucas Mertehikian.
Description: Gainesville : University of Florida Press, 2026. | Includes bibliographical references and index.
Identifiers: LCCN 2025024424 (print) | LCCN 2025024425 (ebook) | ISBN 9781683405559 hardback | ISBN 9781683405696 paperback | ISBN 9781683405917 ebook | ISBN 9781683405788 pdf
Subjects: LCSH: Natural history in art | Natural history—Latin America | Nature—Effect of human beings on—History | Climatic changes in literature | Creation (Literary, artistic, etc.) | Climatic changes—Latin America | Climatic changes—Effect of human beings on
Classification: LCC N6502.6 .I43 2026 (print) | LCC N6502.6 (ebook)
LC record available at https://lccn.loc.gov/2025024424
LC ebook record available at https://lccn.loc.gov/2025024425

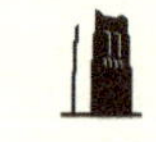

University of Florida Press
2046 NE Waldo Road
Suite 2100
Gainesville, FL 32609
floridapress.org

GPSR EU Authorized Representative: Mare Nostrum Group B.V., Mauritskade 21D, 1091 GC Amsterdam, The Netherlands, gpsr@mare-nostrum.co.uk

CONTENTS

FIGURES

ACKNOWLEDGMENTS

The editors would like to thank the journals that allowed us to publish English versions of the following articles: Gabriel Giorgi, "Desorden del tiempo en el museo anticolonial," *El Matadero* 16, 2022, pp. 27–38; Gisela Heffes, "El necroespacio del Antropoceno: un archivo anacrónico," *Estudios Filológicos,* vol. 72, 2023, pp. 117–134; Ignacio Pastén López and Ignacio Veraguas Caripan, "El museo de la modernidad: archivo fotográfico, colección patrimonial y genocidio Selk'nam en Carlos Gamerro, Eduardo Belgrano y Galo Ghigliotto," *Cuadernos LIRICO* 26, 2024. Special thanks go to the editors of these journals: Patricio Fontana, Alejandra Laera, Julio Premat, and Cecilia Quintrileo.

We would also like to acknowledge the help provided by our home institutions, Georgetown University and Dumbarton Oaks, for making this edited volume possible. We are grateful to Joanna Page for agreeing to write the book's epilogue, which traces connections between the different chapters. Thanks also go to the artists, editors, and filmmakers who allowed us to reproduce their work in the following pages, including Liliana Porter, Leandro Listorti, Emilio Renart's estate, Galo Ghigliotto, and Andrea Palet from Laurel Libros.

Finally, we are grateful to Stephanye Hunter at the University of Florida Press for her enthusiastic support of this project and for her patient guidance.

Introduction

The New Natural History

Nicolás Campisi and Lucas Mertehikian

In Diego Vecchio's *La extinción de las especies* (2017), the Smithsonian's first director, Zacharias Spears (a fictional version of Joseph Henry), faces a challenge: telling our planet's history through an assemblage of very different natural and human-produced objects including embalmed animals, a meteorite recently arrived from Mexico, and an endless series of glass cabinets that he has to fill to draw the public's attention. As he discusses plans with his secretary, Miss Sullivan, to organize the museum's incipient collections, Spears wants to turn away from the logic of the cabinet of curiosities, which, he asserts, sought to entertain audiences rather than educate them. "Mr. Spears's dream," Vecchio writes, "was that whoever visited his museum and was predisposed for such an adventure would embark on a journey to remote spaces and times, traveling in a vehicle much faster than the fastest of railroads, as imagination can be when guarded by science" [El sueño de Mr. Spears era que quien visitara su museo y estuviera predispuesto para tal aventura, emprendiera un viaje hasta espacios y épocas remotas, desplazándose en un vehículo mucho más veloz que el más veloz de los ferrocarriles, como puede llegar a ser la imaginación cuando es custodiada por la ciencia] (35–36). But how to do so remains a mystery. As Spears frantically dictates notes to Miss Sullivan, he realizes that the parsimonious time of planets does not move straightforwardly but rather in zigzags. "Nature never turns back. When presented with an obstacle, it ventures through side paths. Instead of destroying, it prefers to cross out" [La Naturaleza nunca da marcha atrás. Cuando se le presenta un obstáculo, se aventura por senderos laterales. En lugar de destruir, prefiere tachar] (46). Thus, through a series of cabinets containing taxidermied specimens and fossils, the museum comprises geological time and renders life's serendipity at a planetary scale. The rest of the novel explores the different ways in which natural history objects can serve as celebrations of life and reflections on our planet's impending doom, blending scientific theories of evolution with fictional accounts of the origins of life.

Literature is not the only art form that has taken up classical motifs of natural history. In recent decades, visual artists have also resorted to practices and artifacts that traditionally belonged in natural history museums. In 2019 the Museu Nacional da República in Brasília presented a retrospective exhibition of Lia do Rio titled *Tempo em suspensão,* showcasing the artist's long-standing relationship with natural elements that started in the 1990s and continues to this day. Specifically, do Rio's work engages with dried leaves echoing those held in herbaria, plant collections assembled since the sixteenth century by Western botanists traveling across the world and still housed at major scientific institutions. Do Rio's retrospective included sculptures made entirely of dried leaves and tapestries made of patches of leaves sewn together hanging down the museum's walls. Among these works was *O livro das folhas* (1994), a large book whose size and pages resemble those of botanical publications holding valuable information on plants (either glued to the page or drawn by illustrators) such as the name, provenance, and medicinal or edible uses. However, the leaves in *O livro das folhas* take up the whole page, and the artist does not give any information to help the viewer identify the plant's species. Thus, it turns a scientific artifact into an aesthetic device that questions humans' capacity to study, categorize, and therefore master nature.

The claim that brings together the essays in the present book is the conviction that, as these examples begin to show, Latin American writers, artists, and critics have turned to the discourses and practices of natural history as a way of reckoning with the realities of climate change and the Anthropocene as well as with the conceptual and aesthetic challenges that such realities pose to them. As Heather Houser insightfully notes, contemporary artists and writers "follow the classificatory impulse of natural history not out of nostalgic yearning . . . but to show the endurance of naturalism's epistemological and representational traditions" (17). The essays that follow analyze how books, artworks, and contemporary museum practices revisit natural history to explore their aesthetic potential and reconceive the Western ontologies that cast humans and nature as separate entities.

Vecchio's novel and do Rio's work foreground a series of literary and artistic strategies that are at the heart of the books, artworks, and cultural practices that *Imagining a New Natural History: Latin American Cultural Production in the Anthropocene* interrogates. First, writers and cultural agents behave as curators whose task is to organize dispersed objects into a coherent narrative about nature. To do so, they engage with collecting practices traditionally associated with natural history and the museum space. Thus, the characters in these artistic works can be archaeologists, museum curators, botanists, traveling naturalists, and taxidermists, and the privileged genres are the catalog, the cabinet of curiosities, the exhibition, and the collection itself. Furthermore, these characters and narrative devices often defy national and linguistic boundaries, moving across disciplines, historical periods, and geographies. But if art usually deals with the time and space of humans, these newly arranged narratives of nature are also faced with the geological and the planetary. While tracing the human imprint in our planet's deep geological history, these works undo the nature-culture divide upon which modernity was founded. When confronted with these new artistic and literary practices, natural history collections such as anthropological museums and herbaria become explorations into our current climate predicament, making the boundaries between human and planetary scales collapse.

This interest in nature is not entirely new within Latin American culture. There is indeed a long tradition of artistic and literary works that explore humans' relations with their environment. In literature, for example, this spans from nineteenth-century Romanticism to the twentieth-century *novela de la tierra,* which includes some of the region's most renowned novels, such as José Eustasio Rivera's *La vorágine* (1924), Juan Rulfo's *Pedro Páramo* (1950), and Sara Gallardo's *Eisejuaz* (1971). There is also a wide-ranging corpus of works by European naturalists who traveled to Latin America to investigate the region's natural resources, the most salient figure among them being Alexander von Humboldt. In these travel narratives, as in the Romantic and regionalist traditions, Latin America figures as the producer and exporter of raw commodities. As Ericka Beckman has shown, during Latin America's export age, regionalist writers turned their gaze "to the peripheries of the already peripheral nation-state" (158). Instead, contemporary artistic practices are often cosmopolitan and hard to pigeonhole into national and linguistic traditions. Latin America serves as a window to look into processes of planetary destruction and holds accountable past traditions that turned nature into exportable commodities.

There is yet another significant innovation in contemporary engagements with natural history. Many of the works analyzed here take up the point of view of the colonial subject, who was the target of extractive practices. In other words, the collected frequently becomes the collector and the meaning of collecting itself is subverted. In 2019, seventy women of the Munduruku Nation entered the Natural History Museum in Alta Floresta, Brazil, to reclaim more than ten artifacts that belonged to the community. In retelling this story, Eliane Brum explains that to them, these are not archaeological objects but "their ancestors' spirits" that were crying to be taken back to their place of origin (35). In light of anecdotes like this one, we wish to highlight in this volume the ethical urgency of returning to the natural history museum. We conceive this book as an initial exploration of a growing corpus of works that engage with natural history as a discipline and the deep history of our planet as a geological reality that haunts us in times of climate change. The list of works analyzed here is certainly not exhaustive. Still, we believe it is illustrative of a pressing ethical and aesthetic concern with climate change that resorts to discourses and practices often associated with the past. In the following sections of this introduction, we will explore Latin America's centrality in the history of natural history (to borrow John G. T. Anderson's formulation) and how contemporary artists are returning to this discipline to shed light on its political underpinnings and new relevance.

Natural History and Latin America

Natural history as a discipline is comprehensive. Paula Findlen observes that *Historia Naturalis* by Pliny the Elder (ca. 22–78 CE) "offered an expansive definition of the subject" that included "all entities found in nature, or derived from nature, that could be seen in the Roman world and read about in its books; art, artifacts, and peoples as well as animals, plants, and minerals" (437). Very much in the fashion of Zacharias Spears's total museum in Vecchio's *La extinción de las especies,* Findlen describes ancient natural history as an all-encompassing enterprise. Consequently, the practices associated with natural history include naming, collecting, classifying, curating, and displaying. The institutions of natural history are those where these practices flourish, such as the cabinet of curiosities, the botanical garden, the zoo, and the museum. Still, Findlen notes that for Pliny the Elder, "history" underlined the role of description in knowing the natural world (everything that is) rather than "any specific sense of past" (437). In contrast, the books, artworks, and curatorial devices analyzed here examine, dissect, and often subvert the notion of history that comes out of these traditions. They all conceive of it as connected to a sense of past, however remote and simultaneously present.

Moreover, in her definition of natural history as a form of knowledge, Findlen highlights another critical aspect of the works and issues explored in this volume. Unlike physics and astronomy, natural history in the early modern period, drawing from its ancient sources, "reflected a different intellectual model of growth of scientific thought" as it advanced incrementally rather than revolutionarily (437). That is, accumulation plays a major role in natural history, which explains the proliferation of collections and catalogs of nature that were compiled from the sixteenth century onward. This new system of ordering the natural world was shortly after connected with transatlantic exploration and colonial settlement. Thus, natural history became widely associated with the cabinet of curiosities as an exhibition and pedagogical model. Likewise, juxtaposition is a defining aesthetic device of the works that we study here, as it brings together human and nonhuman entities that would normally stay apart.

In his 1966 seminal work *The Order of Things*, Michel Foucault offered a narrower, historically situated definition of natural history when he famously claimed that "the naturalist is the man concerned with the structure of the visible world and its denomination according to characters. Not with life" (176). Foucault was interested in establishing that natural history was a way of knowing specific to the seventeenth and eighteenth centuries. It did not emerge from ancient or medieval versions of "natural history" as broadly defined by Findlen (what Foucault calls "Classical Age"), but rather broke away from them. Specifically, Foucault claims that natural history is the "space opened up in representation by an analysis which is anticipating the possibility of naming" (138). With the publication of *Systema Naturae* in 1735 and *Species Plantarum* in 1753, the Swedish botanist Carl Linnaeus established the binomial nomenclature as a taxonomic principle that would be widely adopted by Western science, allowing Europeans to name (and thus instrumentalize and possess) the natural world.

Although it would prove key to Western expansion, Foucault considered natural history a somewhat poor way of perceiving the world. In order to extract from organisms the features that made them representable and classifiable, naturalists had to forgo all that was not essential to the naming of a species. Thus, sight became the privileged sense of natural history, excluding others such as taste and smell that until the seventeenth century had been relevant to the description of nature. Until naturalists like the Italian Ulisse Aldrovandi (1522–1605), "the history of a living being was that being itself, within the whole semantic network that connected it to the world. . . . Signs were then part of things themselves, whereas in the seventeenth century they become modes of representation" (Foucault 140). The contributors to this book are often interested in how natural history collections, with their lifeless plants and animals, can be reactivated, making signs once again part of the living environment.

Defined either in a broad or a narrow sense, natural history—its institutions and practices—flourished around the same time that Europeans settled in the Americas. The first botanical gardens, for example, were established in Italy and later in the rest of Western Europe, with the foundation of the gardens at Padua, Pisa, and Florence in the 1540s, followed by Montpellier, Leiden, and other universities through the 1630s (Ogilvie 151). Consequently, European gardens and cabinets of curiosities were soon filled with specimens coming from distant and—for them—new geographies. Anthony Grafton, Nancy Siraisi, and April Shelford note that besides providing Europe with gold and other precious minerals, the Americas "also confronted Europeans with new animals, from bison to microorganisms, and new plants, from tobacco to potatoes." The authors find that the revelations "proved both attractive and dangerous enough to change social life—sometimes, indeed, to cause what were perceived as social and cultural crises" (161). Thus, Linnaean taxonomy and collecting practices were turned into the foundations of profitable colonial enterprises at the expense of the knowledge and livelihoods of Indigenous communities.

This trend continued well into the nineteenth century, with large scientific expeditions like the one carried out by Humboldt and Aimé Bonpland that brought back to Europe thousands of plant and animal specimens yet unknown to European science. Often masked as unobtrusive and harmless, categorized by Mary Louise Pratt as "anti-conquest narratives" (*Imperial Eyes* 37), these expeditions were also intimately connected with the commodities trade and certainly filled Europe's natural history collections. Many of the chapters in this book explore the economic, environmental, and racial ramifications of this intense traffic between the Americas and Europe, the outermost expression of which was, of course, the enslavement of Indigenous and African people.

However, despite their violent and complex roots, natural history collections today hold valuable ecological information that can help remediate climate change. In a 2023 special issue of the *Journal of Animal Ecology*, the editors claim that natural history collections can be leveraged to understand the impacts of climate change across the world. "The collections harbored by natural history museums," they write, "span time, taxa, and geographic regions, and, as a result, provide unparalleled opportunities to explore questions in global change biology" (Sanders et al. 232). In zoology, one example of this is that variation in climatic conditions can affect body size. Therefore, the editors write, "specimens from museums provide a wealth of opportunities to explore how phenotypes might change through time" (232). Likewise, dried plants stored in herbaria provide botanists unparalleled opportunities to study environmental change. Herbarium specimens can date back many centuries, allowing the study of "long-term processes such as recent invasions and their genetic population history" (Lang et al. 110). Moreover, some researchers are exploring the possibility of finding viable seeds in herbarium specimens that might eventually allow the regrowing of now-extinct plant species (Marinelli). In these ways, natural history collections are being reactivated by scientists as much as by the writers, scholars, and artists analyzed in this volume.

Contemporary artists and scientists embrace the contradictions of natural history collections, which are riddled with racialized implications but are still the focal point of their work. They return to the natural history archive to unearth information that prompts new research to counter climate change and to look for inspiration for creative works that engage with the environmental crisis. Why, now that the institutions and practices of natural history are deemed obsolete precisely because of their racial and colonial undertones, do so many artists and writers use them as a platform for their aesthetic projects? What draws them to their fascinating yet violent character? Here we take up Donna Haraway's proposal of "staying with the trouble" to sit with this contradiction rather than overcome it. Haraway encourages cultural practitioners to face the damages of colonial histories and reinvent "the conditions for multispecies flourishing . . . in a time of human-propelled mass extinctions" (130). Natural history is appealing precisely because, when seen retrospectively from the standpoint of our climate-change era, it brings to light these fruitful contradictions. It offers practices and technologies that are obsolete yet ripe with possibilities. By tackling them, artists and writers delineate the contours of a new natural history: a cultural practice that seeks to address the aesthetic and political challenges of the Anthropocene.

This book joins numerous critical interventions situated at the intersection of natural history and Latin American cultural studies. For instance, by examining works in multiple aesthetic formats, our book expands the corpus Joanna Page explores in *Decolonial Ecologies,* which encompasses the genres of natural history—including bestiaries, herbaria, and cabinets of curiosities—taken up by contemporary artists. Page finds that artists rework these formats to point out the exclusionary logic of natural history but also to enact a decolonial approach and "construct an alternative modernity that embraces plural ontologies" (16). Page's work is inscribed within a larger academic interest in reëxamining the European expeditions to the New World, as in Mauricio Nieto Olarte's *Remedios para el imperio,* about the "instrumentos de apropiación" (instruments of appropriation) of naturalists and travelers who classified, codified, and exhibited their objects of study (16).

Other scholarly publications in the field include Juan Pimentel and Mark Thurner's edited collection, *New World Objects of Knowledge,* an academic project that expands the reach of collecting and museum practices by providing entries about New World objects that are scattered across the globe and therefore restoring them at least symbolically to their places of origin. Their cabinet of curiosities encompasses animals (Darwin's tortoise and Darwin's hummingbird), codices (the *Codex Mendoza*), and commodities (cacao and rubber), organized in the logic of early modern cabinets instead of contemporary museums. In their words, the book's purpose is "to incite critical curiosity about the New World as a key protagonist in the history of modern knowledge" (2), an objective that many artists examined by the book's contributors share.

Daniela Bleichmar's *Visible Empire* foregrounds the importance of visual epistemology for the ordering and conquering of nature that translated into thousands of images of New World plants and animals; visual culture served as a bridge between science and empire since it was a fundamental part of "producing knowledge and enacting governance" (10). We are interested in joining this vital chorus of voices to highlight how contemporary Latin American artists and writers return to the practices of natural history to politicize them in the context of our climate crisis.

Anthropocene Aesthetics and the Scales of Natural History

Why has natural history become such an urgent focal point for Latin American artists during the Anthropocene? *Peregrino transparente,* a 2023 novel by Colombian writer Juan Cárdenas, follows the steps of the Chorographic Commission led by Italo-Venezuelan cartographer Agustín Codazzi to chart the territories of the Republic of New Granada, which comprised what is currently known as Colombia. The novel focuses on the British painter Henry Price and his discovery of an Indigenous artist called Pandiguando, whose nonrepresentational techniques draw Price's attention to the relationship between realism and racism that characterized the methods of natural history. Told in the first person by a present-day narrator who obsessively reads Colombian expeditionary literature, the novel exemplifies the drive to return to the nineteenth century with the "obscure intuition that everything we are going through today was plotted in that other time" [oscura intuición de que todo lo que estamos atravesando hoy se cocinó en ese otro tiempo] (62). We are interested in authors who recognize that the Anthropocene asks us to go beyond the linear and realist narrative procedures that coincided with the advent of industrialism and natural history. Though using paradigmatic formats of the discipline such as the nature collection and the bestiary, contemporary artists subvert these modern genres to show their implications in the subjection of peoples and their native territories.

Although we use the Anthropocene as the cultural and ecological category that reunites these aesthetic practices, we acknowledge that the label was rejected by the International Union of Geological Sciences (IUGS) in a statement it released in 2024. In it, this group of geologists and geoscientists dismissed the category of the Anthropocene due to its usually assigned chronology, which locates its beginnings in the mid-twentieth century. During this period, which J. R. McNeill and Robert Engelke have called "the Great Acceleration," the first nuclear tests, the explosion of the world's population, and the expansion of a fossil fuel society caused unprecedented impacts in the biosphere (McNeill and

Engelke 4). For the geological union, however, "the Anthropocene has much deeper roots in geological time" (IUGS). While we are attentive to the different histories of the term "Anthropocene," we find that it has been a crucial notion for academics and cultural producers who seek to illuminate the various ecological catastrophes laying waste to contemporary societies.

In the prologue to her multipart novel *Trilogía del agua* (2024), Argentine writer Claudia Aboaf recognizes the inscription of her fictional work in the speculative genres—such as climate fiction, the new weird, and the gothic—which have been making sense of human modification of the natural environment for decades. In this regard, Aboaf is aligned with Amitav Ghosh's argument in *The Great Derangement* about the confined temporal scales of the realist novel and the comparatively expansive reach of speculative fiction, which has explored "the uncanny intimacy of our relationship with the nonhuman" (Ghosh 33). For Aboaf, the Anthropocene is defined by our actions as geological agents who advance the "accumulation of goods, colonial extractivism, displacement of native communities, wars that hide energy demands, and ecocides that are just beginning to be addressed as interspecies genocides" [la acumulación de bienes, al extractivismo colonial, el desplazamiento de comunidades originarias, guerras que esconden demandas energéticas y ecocidios que recién empiezan a ser atendidos como genocidios interespecies] (11). As the umbrella term bringing together these different forms of social and ecological violence, "the Anthropocene" constitutes the horizon upon which contemporary artists and writers reflect on the colonial legacies of natural history.

The Anthropocene designates the capacity of human beings to act as geological agents with a planetary force. Since geological forces such as mountains have taken a life of their own, we now inhabit two conflicting timescales. "In our own awareness of ourselves," remarks Dipesh Chakrabarty, "the 'now' of human history has become entangled with the long 'now' of geological and biological timescales, something that has never happened before in the history of humanity" (7). Unconvinced with the Anthropocene's neglect of capitalism as a terraforming force, Jason W. Moore proposes the term "Capitalocene" to address how social systems can disrupt natural formations (85). Contemporary Latin American culture actively engages with these links between natural history and the region's violent colonial past. Gabriela Wiener's *Huaco retrato* recounts the life of the author's great-grandfather, the Austrian anthropologist Charles Wiener, who stole hundreds of pre-Columbian artifacts from Peru, had a child with a Peruvian woman, and was hailed in France for his pioneering ethnographic work. *Huaco retrato* adopts the format of the cabinet of curiosities to denounce natural history's entanglement with histories of colonial and sexual violence, whose consequences are still felt today.

In addition to questioning the colonial underpinnings of natural history as a discipline, contemporary Latin American artists narrate the Anthropocene by considering the deep histories of extraction and nation-building. Recent critical interventions in the environmental humanities have posited the need to revise our narrative models to make sense of climate change's challenges to the literary imagination. Ghosh has argued that "the climate crisis is also a crisis of culture" because the realist novel has historically focused on human dramas and neglected our daily encounters with nonhuman actors (9). This cultural failure is mainly due to the marginalization of literary genres such as science fiction and fantasy, which Ghosh contends have tapped into the "forces of unthinkable magnitude" that make up the geological history of our planet but have played a subsidiary role in the modern literary system (63). Pratt reminds us in *Planetary Longings* that "the Anthropocene chronotope requires a unique causality in which humans' own actions return to haunt them *in forms they utterly failed to anticipate*" (123, italics in the original). Similarly, Chakrabarty critiques the term "Capitalocene" for narrowly focusing on human agents instead of considering the much longer geological history of the Earth (172). The accounts of European naturalists who domesticated Latin American nature operated under the sign of teleology. Andrea Wulf describes how in the eighteenth century, people began seeing nature as a domain to be improved through cultivation, as "a 'howling wilderness' that had to be conquered" (67). By contrast, these twenty-first-century artworks show that the extractive worldview was the leading cause of the derangements of scale with which the climate crisis has confronted us.

The Anthropocene has prompted a shift from the global to the planetary. Latin American countries, considered at the tail end of modernization and the globalized world, are now among the first to suffer the burdens of human-induced climate change. Latin America has thus emerged as a strategic place to think about the asymmetrical impacts of the climate crisis. If planetarity, Pratt argues, "shifted the focus toward ecological standpoints that conjugated the human with the nonhuman, the living with the nonliving," then Latin America spearheaded the debates about the rights of nature in the face of large-scale extractivism (*Planetary Longings* 11). The principle of Buen Vivir (Good Living) included in the Bolivian and Ecuadorian constitutions promotes the state's responsibility to respect the rights of Pachamama, Mother Nature. The Buen Vivir worldview assembles various Indigenous concepts such as the Aymara *suma qamaña* and the Kichwa *sumak kawsay* that recognize the expanded communities of human and nonhuman beings (Chuji, Rengifo, and Gudynas 111). Planetarity questions the Western concept of progress and the existence of a single universal history dictated by the logic of exploiting bodies and lands. It also rescues the cultural and socioecological traditions of Indigenous peoples whose artifacts were spoils housed and exhibited in European natural history institutions.

The term "Anthropocene" has been criticized for the assumption that all of humanity is equally responsible for the climate crisis. Kathryn Yusoff rescues the "histories of racism" underlying the concept of the Anthropocene, arguing that the division of geologic materials into active and inert was accompanied by the categorization of Black and Indigenous peoples as inhuman (2). This volume's contributors examine natural history's imbrication with histories of colonial violence and slavery, considering Latin America as a fundamental site where non-Western bodies of knowledge contest the extractive practices of white settler nations. Natural history expeditions served the purpose of surveying and instrumentalizing Latin American nature. In contrast, Indigenous and Afro-Latin American knowledge provides new vocabularies for refashioning our relationship with the nonhuman world. For instance, the notion of *sentipensar* (feeling-thinking) used by Afro-Colombian communities on the Caribbean coast challenges the Western division between mind and body, reason and emotion, and life and death (Botero Gómez 302). Similarly, the Indigenous concept of *kawsak sacha* (living forest) designates the gathering of animal, cosmic, mineral, and vegetable beings that contribute to the balance of the rainforest ecosystem (Gualinga 223). Indigenous cosmological thought has become an important archive of nonhuman agency in the face of climate catastrophe, standing against the limited scales of the Western extractive view.

By incorporating these key postdevelopment concepts into their format, contemporary artworks expand traditional notions of authorship and scale to prompt a greater awareness of coexistence with the nonhuman world. Many works studied in this book could be described as geological writing. In *Escrituras geológicas,* Cristina Rivera Garza draws a parallel between the work of contemporary writers, geologists, and paleontologists: the production of deep-time consciousness that situates us in the gap between the distant past that preceded us and the uncertain future that will ensue long after we are gone (13). The main feature of geological writing practices is revisiting artistic and cultural traditions to interrogate the state of things and stand against the myth that the Earth is a tabula rasa, a clean slate. This Anthropocene writing strategy seeks to reanimate the nonhuman world. As Nicolas Bourriaud argues in *Inclusions,* "The artists of the Anthropocene have learned that everything is matter, that nothing can constitute an undifferentiated and inert background onto which an activity is affixed" (95). In this sense, these Latin American artworks invoke the principal tenets of natural history but bring together spheres formerly conceived as separate, such as the animal and the mineral, showing our interdependence with the more-than-human world.

Instead of perpetuating the exclusionary logic of Western modernity, Latin American artworks delineate the artist's figure as a medium that communicates with the natural and spiritual worlds. This inclusive view of the creative process stems from landmark works of Indigenous thought, such as Davi Kopenawa's in *The Falling Sky*, written with Bruce Albert. Against the backdrop of continuous depredation of the Amazon rainforest, Kopenawa speaks about the importance of dreaming the forest and listening to the spirit beings suffering from the actions of extractive cannibals and earth eaters. "The spirits made me understand that the forest was not endless," Kopenawa observes, "as I once thought it was" (257). Kopenawa inverts the colonial logic promoted by several European naturalists who designated the Indigenous populations of the Amazon as savages, referring to extractive projects as contemporary forms of cannibalism that fill the environment with epidemic fumes. Other Indigenous writers express an affective relationship with the land, in what Ailton Krenak calls "affective alliances" (44), by envisioning the spirits attached to earth beings and the existence of alternative practices of environmental world-making. Inspired by non-Western ways of inhabiting the natural world, various Latin American artists and writers conceive their work through multispecies anthropology, communicating with invisible forces or navigating the vast scales of our planet's history.

When considering the influence of Indigenous world-making practices in contemporary artworks, we must also address how artists reproduce the communal principles of postdevelopment economies. In the face of the Anthropocene's "passive present," whose consequences seem "beyond our reach to cancel," Latin American artists promote active ways of engaging with cultural artifacts (Danowski and Viveiros de Castro 5). Among these principles, writers like Rivera Garza have highlighted the Indigenous concept of communality as constitutive of the creative practices that challenge capitalist and environmental violence. Stemming from the Mixe Indigenous peoples of Oaxaca, communality designates the ethics of reciprocity linked to the practice of *tequio*, the collective unpaid labor each community member must perform for the greater good. In *The Restless Dead*, Rivera Garza differentiates communality from strategies of "giving voice to the voiceless," recognizing that the words of others are equally important as one's own (5). "In unveiling work created by many people in community," Rivera Garza remarks, "communalities of writing address survival strategies based on mutual care and the protection of the common good, challenging the ease and apparent immanence that marks the languages of globalized capitalism" (5). By acknowledging the voices of human and nonhuman actors in times of unprecedented changes to the geosphere, the artists whose work we examine posit new ways of navigating the disparate legacies of Latin America's natural history.

Chapter Outline

The book is divided into four sections that reveal the strong purchase of natural history discourses in contemporary Latin America. The Age of Discovery had profound and ongoing ecological effects in the Americas since many voyages of exploration were part of imperial enterprises that sought to map territories and incorporate them as colonial domains. Still, as Anderson points out, "it is hard to think of any natural historian of the nineteenth or twentieth centuries who was not also a conservationist in some form or other" (256). The four sections of this book grapple with this contradiction and elucidate how contemporary artists and writers return to the colonial era and the nineteenth century with the certainty that many of these practices have steered our planet to the verge of ecological collapse.

The first section, "The Institutions of Natural History," presents essays that center literary, artistic, and curatorial practices within different spaces and devices of knowledge production in natural history—namely, the ethnographic museum, the human zoo, the natural history museum, and the cabinet of curiosities. The contributions in this section engage with the nineteenth-century universal exhibitions as represented in recent Argentine and Chilean literature, pedagogic and curatorial narratives in Mexico City museums, and contemporary artworks that adopt the form of the *Wunderkammer* or cabinet of curiosities to reconfigure natural history from an aesthetic lens. Thus, this first section introduces one of the central claims that structure this volume: the conviction that natural history is embedded within a set of historically well-defined institutions, practices, and discourses, and is also a dynamic framework for examining how humans and nature are inextricably intertwined.

In the first chapter, Ignacio Pastén López and Ignacio Veraguas Caripan explore the dark underside of the Enlightenment project symbolized by the nineteenth-century universal expositions, especially the Great Exhibition in London's Crystal Palace (1851) and the 1889 Exposition Universelle in Paris. Through the analysis of three contemporary novels by Argentine and Chilean authors—*Fuegia* by Eduardo Belgrano Rawson, *El museo de la bruma* by Galo Ghigliotto, and *La jaula de los onas* by Carlos Gamerro—Pastén López and Veraguas Caripan foreground the natural history tradition of human zoos: the deliberate imprisonment of Indigenous peoples, in this case the Patagonian Selk'nam, whom European audiences viewed and consumed as the primitive other. Instead of reproducing the myth of transparency and triumph over nature emblematized by the world's fairs, these three authors resort to documentary writing practices to show that the appearance of Patagonia in the universal expositions coincided with the disappearance (or, more appropriately, the genocide) of its Native populations.

Emily Hind's chapter examines the twenty-first-century combination of natural history and children's museums. Specifically, Hind analyzes the Papalote Museo del Niño, a children's museum in Mexico City's Chapultepec Forest. Through a LEGO city, a tree decorated with flora and fauna, and a video game about an axolotl, children are invited to experiment and design solutions to the city's multiple environmental problems (mainly air pollution, water shortages, and wildfires) within the museum's safe space. Hind explores the diverse artistic and literary traditions at Papalote, such as the extensive corpus of texts on the axolotl. She also underscores the limitations of the museum's initial task to overturn the Anthropocene through its sensorial connection with the cityscape: the smell of smog, the noise of traffic, and the views of a polluted neighborhood.

In his chapter, Jerónimo Duarte-Riascos studies Liliana Porter's 2011 installation *El hombre con el hacha y otras situaciones breves*. He argues that the piece oscillates between the presentation and the representation of the real and in doing so, reveals many of the improbable certainties that determine our relation to others and to the world. Underscoring this improbability, and perhaps paradoxically, the installation functions in a manner not unlike that of historical cabinets of curiosities, deterritorializing spectators and taking them to another time and place. Yet, he claims, Porter's contemporary cabinet of curiosities collapses our certainties and promises a revelation that is always already deferred, in a nod to Jorge Luis Borges's understanding of the aesthetic act as the imminence of a revelation.

The second section, "Natural History and Ancestral Knowledge," builds on the previous chapters to explore the knowledge of human and nature relationships produced by communities that were the subject of colonial violence often justified through natural history's mission. The essays in this section examine Indigenous knowledge and practices forcefully erased by Western understandings of nature, in order to either recover them from the archives of natural history or to analyze contemporary cultural productions that emerge from renewed contact with such Indigenous histories and epistemologies. Significantly, the communities that were often the object of inquiry of natural history institutions here become the subjects of challenging narratives that seek to broaden and enrich our understanding of the living environment, an aesthetic and ethical statement in times of climate crisis.

Luciana Martins examines the biocultural archive of British botanist Richard Spruce (1817–1893), who traveled across the Andes and the Amazon and collected Indigenous artifacts and plant species. She foregrounds the *caraipé* bark that Indigenous communities used to make fireproof pottery. Martins demonstrates that carefully examining the various uses of this plant offers a different vision of the Amazon rainforest. This biocultural archive disarticulates the European image of the Amazon as a green hell; instead, the rainforest becomes the site of countless entanglements between human and nonhuman beings that had remained muted in the archives. Martins invites researchers to retrace the itineraries of these materials from other perspectives, such as those of the Indigenous peoples who accompanied explorers in their travels. This reimagining, she argues, could help trace how landscapes were altered by European expeditions and how these extractive practices are continued today. But the reimagining can also point to other ways of conceiving the past and the future.

Florencia Garramuño focuses on the work of Mapuche artist Seba Calfuqueo and Krenak philosopher Ailton Krenak. Calfuqueo assembles artworks that lay bare the consequences of the 1981 Water Code promulgated by the Augusto Pinochet dictatorship (1973–1990), which sought to render all water marketable. The artist contrasts the notion of water as a commodity with the ancestral conception of this vital resource as a living being. In his texts that bridge the gap between written and oral cultures, Krenak also considers water as a living entity (our grandfather) currently under attack. Garramuño argues that by producing texts that move between different aesthetic formats and considering water as a life form in continuity with the human figure, these artists conceive of a "poetics of an expanded humanity." Garramuño elucidates the work of artists who bridge the traditional gap between nature and culture and in doing so, generate new epistemologies based on ancestral knowledge systems.

In her chapter, Matylda Figlerowicz analyzes two novels written by Indigenous authors, *El infierno del paraíso* by Nahua writer Crispín Amador Ramírez and *Woman of Light* by Chicana writer Kali Fajardo-Anstine. She shows how Indigenous storytelling traditions challenge the human-centered conception of the environment that permeates natural history's scientific discourses and practices. On the one hand, Figlerowicz intervenes in world literature debates that link the novel form to European modernity and the dramas of the human. On the other, she contends that Indigenous narratives can foreground other epistemological traditions that decenter the hierarchies of Western science. The contemporary Indigenous novel, Figlerowicz argues, becomes a porous narrative site where it is possible to imagine new ways of conceiving the relationship between organic and inorganic matter, between the human and nonhuman.

Chapters in the following section, "Ecocriticism, New Materialism, Posthumanism," track the transformations of artistic form considering the planetary time of the Anthropocene. While tracing connections with earlier works such as the jungle novels of Alejo Carpentier and the *indigenista* novels of José María Arguedas, the articles in this section point to the deep timescales of contemporary artworks that explore the use of waterscapes as dumping grounds for dissident bodies, the end of our traditional conception of the landscape when we take into account the ruins left behind by modernizing projects, and the end of the myth of exceptionalism when we consider human beings as living matter in symbiotic continuity with the rest of the cosmos. Each chapter charts the parallel redefinition of the artwork and the novel as aesthetic devices capable of holding more-than-human temporalities and the natural history museum as a space that can accommodate non-Western conceptions of the environment during our age of proliferating ecocides.

Gisela Heffes theorizes what she calls "the necrospace of the Anthropocene," an anachronistic archive of organic and inorganic matter that reveals the different erasures hidden in the continent's official histories. She analyzes three works that center water as a counterhegemonic archive: the short stories "Los pescadores de vigas" by Horacio Quiroga and "Agua" by José María Arguedas and the novel *El Rey del Agua* by Claudia Aboaf. Heffes argues that these narratives embed our planet's deep pasts and futures and turn necropolitical violence such as extractivism and dictatorial violence into collaborative alliances that allow us to build a different image of history. She shows how Latin American writers assemble fluid landscapes that reflect the multidirectional time of the Anthropocene.

In his chapter, Carlos Fonseca theorizes the contemporary ecological novel, which he associates with works that lay bare the violence of history embedded in the landscape. Building on Rivera Garza's notion of geological writing, Fonseca proposes the concept of land archive to designate the vision of the environment as traversed by multiple histories of ecological and political violence. He shows how contemporary writers like Edmundo Paz Soldán, Pola Oloixarac, and Rita Indiana continue the legacy of Alejo Carpentier and the *novela de la tierra,* which overturns the ecological fantasies of Latin America as a zone of endless extraction. Instead, Fonseca observes, these writers confront us with the end of nature and the need to reconstruct the ruined landscape that Latin America has become after the end of history and the modern dreams of infinite extraction.

In his chapter, Gabriel Giorgi analyzes Galo Ghigliotto's *El museo de la bruma* as an example of the changes in the narrative space caused by the new pressure of large-scale temporalities and nonhuman agents. Giorgi foregrounds how the Chilean author captures the surviving traces of the colonization of Tierra del Fuego that expanded throughout the twentieth century. The museum form allows Ghigliotto to juxtapose temporalities that cannot be reduced to any narrative synthesis; he showcases the complicity between museum, exhibition, and coloniality. Giorgi argues that the narrative laboratory of Ghigliotto's novel illuminates the question of territory as a central issue of the planetary imaginaries of our present.

In this section's last chapter, Florencia Malbrán explores how artistic practices can reconfigure our image of the cosmos and the web of life. These two concepts draw from natural history traditions and are central to environmental thinking in the Anthropocene. Malbrán delves into the work of Argentine artist Emilio Renart, who in the 1960s anticipated current debates about posthumanism and the nature-culture continuum. Specifically, Malbrán focuses on Renart's idea of "integralism," which sought to build a symbiotic relationship between man and the cosmos through the combination of various media, painting, sculpture, and drawing. In a series of works entitled *Bio-Cosmos*, Renart sought to undo Western dualisms to unite the different species inhabiting our planet. Malbrán argues that Renart was a pioneering artist because he introduced a new sense of planetarity that emerges when humanity is seen as living matter rather than the measure of all things.

The book's last section, "Human and Nonhuman Histories," continues the reflection on the presence of animals, plants, and minerals in contemporary artworks and museum spaces. The cultural materials analyzed in this section comprise artistic exhibitions about natural phenomena such as hurricanes, documentary films about herbaria and plant collecting, and imagined bestiaries and encyclopedias that reconceptualize the classificatory methods of natural history. What brings together these disparate objects is the artists' need to reframe the narrative mechanisms of natural history in the face of overwhelming climatic events or underwhelming conservation efforts on the part of state and private institutions. The authors and curators of these cultural productions emphasize the materiality of media or play with genres to reflect on the participation of artworks in the conservation of nonhuman nature in times of both massive and invisible extinctions.

Valeria Meiller argues that contemporary art exhibitions produced in recent years that address climate change are deeply connected with academic, political, and activist discourses on the same issues. In doing so, she not only discusses the work of Latin American scholars and artists but also adopts a hemispheric perspective to investigate North-South dynamics and how these affect artistic practices addressing environmentally induced trauma. Specifically, Meiller studies the exhibitions *Simbología: Prácticas artísticas en un planeta en emergencia* (Centro Cultural Kirchner, 2021–2022), *El Dorado* (various sites, 2023–2024), *no existe un mundo posthuracán* (Whitney Museum, 2021–2022), and *Life between Walls* (MoMA PS1, 2022–2023). Meiller finds that art is not only a site of mourning and visibility of ecological trauma, but one that pushes the boundaries of aesthetic practices and might be able to redefine our relationship with the past and present, codified by natural history accounts and collections.

In his chapter, Antonio Gómez posits the figure of the artist as a collector through an analysis of the documentaries *Homo Botanicus* by Guillermo Quintero and *Herbaria* by Leandro Listorti. These films chronicle the private initiative of individuals who seek to compile objects, stories, and natural specimens to illuminate counterhegemonic histories and compensate for the lack of official institutions that would fulfill this public need. While Quintero follows two botanists who seek to preserve a forest's species in the face of rapid environmental change, Listorti draws parallels between the conservation of plants and old films to criticize the precarious state of government-funded conservation of culture and nature today. Gómez shows how these artists revitalize the eighteenth-century tradition of encyclopedism in collections that, even when they are arbitrary and contingent, delineate Latin America as a locus of conservation rather than a site of extraction.

Finally, Nathaniel Wolfson explores contemporary poetry's forms of convivial life among human and nonhuman species. He analyzes Maria Esther Maciel's *Pequena enciclopédia de seres comuns,* a fictional encyclopedia that references ancient naturalist manuals. In Maciel's project, the reader encounters a rhetorical "commons" that sustains hybrid relationships between species. Fascinated by onomastics, Maciel explores the function of proper names like Maria and João in definitional entries of plant and animal beings. She uses methods of appropriation of natural history texts and botanical and zoological documents to play with encyclopedias' taxonomic and semiotic conditions. Despite the diversity of the species she registers, each unique being is, in her invented encyclopedia, poetically related to other beings because of shared aspects of their names. Drawing from Maciel's previous books such as *O livro de Zenóbia,* Wolfson reads *Pequena enciclopédia* as a poetic work sharing certain tendencies with the traditions of conceptual poetry as well as a hybrid literary object imitating historical encyclopedias and natural history documents.

As might be gathered from the preceding summaries, natural history has been experiencing a renaissance. We can only expect this trend to continue in the coming years. During the Covid-19 pandemic, people took to nature and widely used cellphone apps like Merlin Bird ID to learn about the bird species they sighted and PictureThis to discover the names of plants. People also recorded sightings of animals that had not been seen in urban environments in a long while. Time will tell if these serendipitous encounters will be reflected in future literature and artworks that further engage with natural history collections and practices. In Latin America, there are many examples of this upsurge of natural history narratives produced during a time of generalized lockdowns.

The artists and writers we study participate in this revival of natural history. They document endangered animal and plant species ahead of possible mass extinctions or return to historical periods when natural historians were recognizing the dangers of irreversibly altering the environment. Many artists highlight natural history's links with forms of ecological imperialism; others turn our attention to the connections between dictatorship and neoliberalism, the encroachment on Indigenous lands, and the construction of massive infrastructures that have displaced human and nonhuman populations. The artists we analyze in this book all tell what Rachel Carson called "fables for tomorrow," stories about the dangers of modifying nature and threatening our collective survival as a species.

Works Cited

Aboaf, Claudia. *Trilogía del agua*. Alfaguara, 2024.

Anderson, John G. T. *Deep Things out of Darkness: A History of Natural History*. University of California Press, 2013.

Beckman, Ericka. *Capital Fictions: The Literature of Latin America's Export Age*. University of Minnesota Press, 2013.

Bleichmar, Daniela. *Visible Empire: Botanical Expeditions and Visual Culture in the Hispanic Enlightenment*. University of Chicago Press, 2012.

Botero Gómez, Patricia. "Sentipensar." In *Pluriverse: A Post-Development Dictionary*, edited by Ashish Kothari, Ariel Salleh, Arturo Escobar, Federico Demaria, and Alberto Acosta. 302–305. Tulika, 2019.

Bourriaud, Nicolas. *Inclusions: Aesthetics of the Capitalocene*. Sternberg, 2022.

Brum, Eliane. *Banzeiro Òkòtó: The Amazon as the Center of the World*. Translated by Diane Whitty. Graywolf, 2023.

Cárdenas, Juan. *Peregrino transparente*. Sigilo, 2023.

Chakrabarty, Dipesh. *The Climate of History in a Planetary Age*. University of Chicago Press, 2021.

Chuji, Mónica, Grimaldo Rengifo, and Eduardo Gudynas. "Buen vivir." In *Pluriverse: A Post-Development Dictionary,* edited by Ashish Kothari, Ariel Salleh, Arturo Escobar, Federico Demaria, and Alberto Acosta. 111–114. Tulika, 2019.

Danowski, Déborah, and Eduardo Viveiros de Castro. *The Ends of the World.* Translated by Rodrigo Nunes. Polity, 2017.

Findlen, Paula. "Natural History." In *The Cambridge History of Science,* volume 3, *Early Modern Science,* edited by Lorraine Daston and Katharine Park, 435–468. Cambridge University Press, 2006.

Foucault, Michel. *The Order of Things: An Archaeology of the Human Sciences.* Routledge, 2002.

Grafton, Anthony, Nancy Siraisi, and April Shelford. *New Worlds, Ancient Texts: The Power of Tradition and the Shock of Discovery.* Harvard University Press, 1995.

Ghosh, Amitav. *The Great Derangement: Climate Change and the Unthinkable.* University of Chicago Press, 2016.

Gualinga, Patricia. "*Kawsak Sacha.*" In *Pluriverse: A Post-Development Dictionary,* edited by Ashish Kothari, Ariel Salleh, Arturo Escobar, Federico Demaria, and Alberto Acosta. 223–226. Tulika, 2019.

Haraway, Donna J. *Staying with the Trouble: Making Kin in the Chthulucene.* Duke University Press, 2016.

Houser, Heather. *Infowhelm: Environmental Art and Literature in an Age of Data.* Columbia University Press, 2020.

International Union of Geological Sciences (IUGS). "The Anthropocene." Statement, March 20, 2024.

Kopenawa, Davi, and Bruce Albert. *The Falling Sky: Words of a Yanomami Shaman.* Translated by Nicholas Elliott and Alison Dundy. Harvard University Press, 2013.

Krenak, Ailton. *Ancestral Future.* Translated by Alex Brostoff and Jamille Pinheiro Dias. Polity, 2024.

Lang, Patricia L. M., Franziska M. Willems, J. F. Scheepens, Hernán A. Burbano, and Oliver Bossdorf. "Using Herbaria to Study Global Environmental Change." *New Phytology* 221 (2019): 110–122.

Marinelli, Janet. "Back from the Dead: New Hope for Resurrecting Extinct Plants." *Yale Environment 360,* July 12, 2023. https://e360.yale.edu/features/plant-de-extinctions-herbariums.

McNeill, J. R., and Peter Engelke. *The Great Acceleration: An Environmental History of the Anthropocene since 1945.* Harvard University Press, 2014.

Moore, Jason W. *Capitalism in the Web of Life: Ecology and the Accumulation of Capital.* Verso, 2015.

Nieto Olarte, Mauricio. *Remedios para el imperio: Historia natural y la apropiación del Nuevo Mundo.* Uniandes, 2006.

Ogilvie, Brian. *The Science of Describing: Natural History in Renaissance Europe.* Chicago University Press, 2008.

Page, Joanna. *Decolonial Ecologies: The Reinvention of Natural History in Latin American Art.* Open Book, 2023.

Pratt, Mary Louise. *Imperial Eyes: Travel Writing and Transculturation.* Routledge, 2008.

Pratt, Mary Louise. *Planetary Longings.* Duke University Press, 2022.

Rivera Garza, Cristina. *Escrituras geológicas*. Iberoamericana, 2022.

Rivera Garza, Cristina. *The Restless Dead: Necrowriting and Disappropriation*. Translated by Robin Myers. Vanderbilt University Press, 2020.

Sanders, Nathan J., Natalie Cooper, Alison R. Davis Rabosky, and David J. Gibson. "Leveraging Natural History Collections to Understand the Impacts of Global Change." *Journal of Animal Ecology* 92 (2023): 232–236.

Thurner, Mark, and Juan Pimentel, eds. *New World Objects of Knowledge: A Cabinet of Curiosities*. University of London Press, 2021.

Vecchio, Diego. *La extinción de las especies*. Anagrama, 2017.

Wiener, Gabriela. *Huaco retrato*. Literatura Random House, 2021.

Wulf, Andrea. *The Invention of Nature: Alexander von Humboldt's New World*. Vintage, 2015.

Yusoff, Kathryn. *A Billion Black Anthropocenes or None*. University of Minnesota Press, 2018.

I

The Institutions of Natural History

1

The Museum of Modernity

Photographic Archive, Patrimonial Collection, and Selk'nam Genocide

Ignacio Pastén López and Ignacio Veraguas Caripan

Translated by Nicolás Campisi

The triumph of transparency in the modern era can be broken down into a series of architectural landmarks whose similarity lies in the abandonment of monumental heaviness in favor of the lightness of structure and glass. Such buildings have sought to illuminate and exhibit their interiors, a purpose that has allegorically mobilized transparency through construction materials and the representational logic of the architectural composition.[1] Not only did the walls become transparent, but these architectural projects also longed for that pristine quality in a search for sincerity or self-evidence. In this context, it is not a minor fact that thinkers from different latitudes point to the 1889 Universal Exposition of Paris as a prolegomenon of global modernity in the key of extractivist capitalism, celebrated in works that heralded the glory of transparency. The exhibition of the Indigenous peoples of Patagonia in the Universal Exposition laid the groundwork for the future expropriation of their lands and the arrival of several extractive industries such as logging and salmon farming that continue to plunder the region to this day. Roland Barthes points to the regional history of the Universal Exposition as a founding enclave of industrial reason coupled with a way of understanding a museum aesthetic favorable to global markets, typical of the metallurgical framework of the then newly inaugurated Eiffel Tower (34).

More recently, Peter Sloterdijk ponders the Great Exhibition of 1851 held in London's Crystal Palace, a precedent of the Universal Exposition, as a metaphor for globalization, marked by air-conditioning and comfort in the immanence of purchasing power (55). In the context of the Great Exhibition, the Crystal Palace was an interior space that boasted its transparency and sought the complete absorption of the exterior; it wished to generate a universe with the totality of the known world in the climatic womb that sheltered its skeleton. The transition from the Great Exhibition to the Universal Exposition marks a gradual desire for the transparency of the world, the vision of the immanence of capital. Sloterdijk develops this metaphor based on Fyodor Dostoyevsky's *Notes from Underground* (1864), which already allows us to trace the connection between narrative and the legibility of the world.

However, the triumph of transparency and the clear-cut amplitude flaunted by globalist civilization has its opacities and mists. In the context of Latin Americanism, Alejandra Uslenghi has highlighted the importance of the Argentine and Chilean pavilions of the Parisian Universal Exposition as symbols of the forced introduction of the colonies into the cultural and governmental hegemony of the old continent's empires (57).[2] For their part, Solène Bergot and María José Correa have studied the staging of modernity in the national expositions held in Chile, in both their educational pretensions and nationalist celebrations.[3] During the past few decades, literary-historical works written by Argentine and Chilean authors have taken up the Universal Exposition of Paris as a point of departure to map other figurations and representations of transatlantic exchange and natural history.

In this article, we examine how the novels *Fuegia* (1991) and *La jaula de los onas* (2021) by Argentine writers Eduardo Belgrano Rawson and Carlos Gamerro, as well as the narrative artifact *El museo de la bruma* (2019) by Chilean writer Galo Ghigliotto, have disputed the way the Latin American pavilions in Paris represented the region's insertion into modernity. Instead of reinforcing the myth of Latin America's arrival at the banquet of the world, the books show that the cosmopolitanism of the Southern Cone is only possible under the premise of the extermination of the Fuegian peoples. Such sacrificial genocide dismantles the curtain that covers the Indigenous human museum, as the point of convergence of the books is the exhibition of state archives regarding the massacre. Using archival resources typical of contemporary documentary narratives, such as patrimonial collections, photographic records, and survivors' testimonies, the novels underpin the truncated modernity of the Universal Exposition as the starting point for the policies of extermination of the peoples of Tierra del Fuego, representing the last voices who could narrate this horrific episode that took place amid the harsh Patagonian landscape.

The architectural precedents of the Universal Exposition are to be found in the Victorian-style Palm House, a greenhouse located in the Royal Botanic Gardens at Kew (1844–1848) and designed by the architect Decimus Burton and the engineer Richard Turner. This building was the first large-scale construction with cast iron in England, and its graceful body laid the foundations for the modern obsession with transparency. The greenhouse was the precedent of the Crystal Palace, commissioned to the architect Joseph Paxton for the Great Exhibition of 1851. The Universal Exposition of Paris continued this architectural model and contributed to the first significant constructions that included air and emptiness as organizational principles; transparency stands as an epic of dematerialization (Artiles Burgos and Boned Purkiss 50). Moreover, these architectural impulses of modernity were later underlined in Modern Architecture, an international exhibition held at MoMA in 1932.

It is both curious and macabre that the pavilions that yearned to include the known world were preceded in constructive terms by greenhouses, even more so when we recall that the interior of the Great Exhibition contained "living specimens" in human zoos, the public display of Indigenous people in a so-called primitive state. After housing climate-controlled organic life in their previous iteration as greenhouses, these buildings then served as mechanisms of global (climate) control through the forceful imprisonment of people in cages. Throughout the nineteenth century, several Indigenous people from Patagonia were exhibited in human zoos as spectacles of exoticism and purported Indian savagery. Before the Universal Exposition displayed members of the Selk'nam community, eleven Fuegians in 1881 and several Mapuche in 1883 became objects of similar ethnographic exhibitions (Peña and Zurita 15).

The triumph of transparency is now the emblem of a corporate and international building style. The precedent of glass has shifted from pristine interiors to reflective blocks since, in urban buildings, the windowpanes mirror the still-standing historical hulks of the city or the passing clouds. Concerning the prevalence of transparency in contemporary architecture, Hal Foster asks, "Is this space civic or touristic, one of social use or mass distraction—or is the distinction now quite blurred?" (43). At the turn of the twenty-first century, transparency pretended to show itself as such, again, in the Volkswagen car factory in Dresden, Germany, the so-called Transparent Factory (*Gläserne Manufaktur*) inaugurated in 2001. In a metaphorical reading of the formal associations produced by the Transparent Factory, Octavi Comeron has proposed a simile with the museum. Museum spaces, especially since the 1960s, have tried to lighten the weight of their walls, thus opening the institution to the city's social space, giving visibility to what they contain, or even showing the museum's inner workings (Comeron 32).

It is not unreasonable to situate the Universal Exposition in chronological and metaphorical terms in the middle ground between the greenhouse and the factory. The constructive logic varies from adequate air-conditioning for organic growth to the grouping of images of the hemispheres and the exposure of workers' operations. The creation of an interior that pretends to absorb the totality of the external world, or the entrance of the universal into the interior forged by a glass receptacle, equates life on display with the pieces of technical development as the foundational elements of an extractivist model. Unlike the greenhouse, the glass interior of universal exhibitions was not an adequate construction for the growth of life but instead displayed the adequacy of life for the development of transnational markets. As Jean-François Lyotard has asserted, the libidinal logic of the interior is understood as an imaginary without a symbolic register (22). However, the principle of transparency stands in opposition to the opacity surrounding the histories of humans who were exhibited within these buildings. Archival sources appear as alternative materials to glass and iron, with which European knowledge sought to construct universality.

By contextualizing these three Southern Cone novels within this more extensive cultural history, we find that they do not simply surrender to the logic of transparency as an emblem of power and civilization. It is not transparency, understood in both spatial and temporal terms, that we find in these three literary works but on the contrary, the mist of facts, stories, and images of the Patagonian past. In each novel, the proliferation of archival sources documenting the Indigenous genocide opposes the triumph of transparency in modern architectural interiors. By structuring their novels as opaque architectural spaces, the three authors show how the obsession with transparency served in equal terms to hide and display the foundational violence of Western modernity, especially the subjection of Indigenous peoples who were exhibited as live specimens in universal pavilions. While structuring their novels through the format of an archive or a museum, the writers reflect on how these modern cultural institutions—and their predecessors such as cabinets of curiosities and human zoos—played a crucial role in the accumulation by dispossession of Indigenous bodies and territories in the southernmost part of the world.

Album: *Fuegia*

Fuegia, the third novel by the Argentine writer Eduardo Belgrano Rawson, introduces us to the first forms of representation of Latin American citizenship through the lens of Western culture beyond the Atlantic. The representational struggle in the novel condenses the maxim that the Chilean writer Ronald Kay had already stated in *Del espacio de acá* (1980), that the appearance of photography in the second half of the nineteenth century, propitiated by the advances of the Industrial Revolution and its landmark monuments such as the Eiffel Tower, relegated the nations of the Southern Cone, which still faced artistic dichotomies typical of iconoclasm, to peripheral status (38).[4] *Fuegia* tells the story of a liminal and sacrificial people condemned to disappear (Pratt 22). Placing the compass in Tierra del Fuego, the third-person voice tells of the sorrows of the Selk'nam (Ona) and Yámana (Yagán) Indigenous peoples, whose forms of existence were exterminated in the genocides that propitiated the economic opening of the channels and ports of the Argentine and Chilean extreme south, especially thanks to the exploitation of gold, the plantations for sheep breeding, and the hunting of whales and sea lions.

The peculiarity of the novel is that it is presented as docufiction, a hybrid genre in which documentary and fictional elements are mixed (von Tschilschke and Schmelzer 16). Or, better yet, in the words of Magdalena Perkowska, *Fuegia* appears as a visual fold that knots and dismantles photographic narratives about the Fuegians (138). The novel incorporates thirteen photographic images: a montage on the cover and a photograph at the beginning of each of its twelve chapters. The text also refers to five images that allude to common uses of photography during the period portrayed in the novel: ethnographic photography, private photography of affluent people, and postmortem photography (Perkowska 136).

The novel tells the story of an Indigenous family, the couple Camilena and Tatesh and their three children, who inhabit the threshold of two times and spaces. First, Camilena and Tatesh are from two different Indigenous lineages. She represents the Yámana, the canoe people of the south, who are called *canaleses* (channelers) in the novel; he belongs to the Selk'nam—called Parrikens in Fuegia—of the northern steppes, a people of hunters. Second, the family journeys from the south to the north, from the Anglican mission where we meet them at the novel's beginning, to the northern ranches where they disappear at the end. The family's flight from south to north due to the harsh conditions that decimate the canoeists in the south turns out to be a descent into hell in the annihilation of the north by the settlers. The episodes of these three plots—the South, the voyage, and the North—do not form a linear narrative but are a series of scenes organized around certain groups of characters from whose perspective the events are told.

The novel can be categorized as documentary fiction, a photographic album of the massacre of the expatriated southern Patagonian communities, particularly that of Fuegia Basket, a nine-year-old Yámana girl who, in the company of three other members of her ethnic group, was kidnapped and sent to London to become civilized and satisfy the curiosity of the British, as well as to attend the universal exhibitions. It is to these four Fuegians—Fuegia Basket, Jemmy Button, York Minster, and Boat Memory—that Belgrano Rawson dedicates his novel, narrating the story of their genocide. Although Belgrano Rawson has insisted that *Fuegia* is not a historical novel, we can indeed note that the novel functions as a counterarchive of the stories narrated in documentary texts of the time, such as José María Borrero's *La Patagonia trágica* (1928) and Roberto Payró's *La Australia argentina* (1898) or, with greater emphasis, the visual account of the images in the photographic album of Julius Popper, whose 1886 scientific expedition was one of the first attempts to colonize Tierra del Fuego (Tello 21). As Sonia Jostic rightly points out, the Romanian's expedition, whose purpose was to exploit gold deposits with the support of the Buenos Aires government, culminated in the violent expulsion of the communities that occupied the territories delimited by Tierra del Fuego.

In addition to being an explorer and engineer, Popper was a fan of military photography; one photograph records him and his personal army surrounded by corpses that had been shot during an "Indian hunt" [cacería de indios] (Belgrano Rawson 103). If Popper's photograph represents white adults murdering natives, the photographs introduced by Belgrano Rawson are primarily photographs of infants, even newborns, which leads Jostic to suggest that photographs should "be thought of as a corpus, as meaningful assemblages, and not as universes of meaning that are exhausted in each image" [como un corpus, como conjuntos significativos y no como universos de sentido que se agotan en cada imagen] (279). Such meaningful assemblages evoke a particular register: the genealogies that each of the photographed children reaches for throughout his or her life in the conquered territories.

The cover photograph, the figure of a baby carried on the back of an Indigenous woman, is particularly relevant in this retrospective examination of the visual compendium of the tragic fate of Fuegia's infants. The caption accompanying this image in the text explains that the baby is Camilena and the woman is her mother. In Perkowska's view, the child "represents the future of the generation personified in the woman, the future that will be truncated, because, according to the text, both adults and children will be annihilated by the expansion of Western culture and economy" [representa el futuro de la generación personificada en la mujer, el futuro que será truncado, porque, de acuerdo con el texto, tanto los adultos como los niños serán anihilados por la expansión de la cultura y economía occidentales] (147). The twelve photographs, therefore, are portraits displayed in a sort of ornamental frame. Each depicts a child and his or her family lineage, being eight European, two Indigenous, and two mestizos. *Fuegia,* contrary to the approaches of Annette Kuhn or Marianne Hirsch, who see the family gaze of photographic albums as a condensation of experience in favor of a private archive cared for by the descendants, cuts the photographic experience in a parallax.[5]

If children born in the West can imagine family lives that memorialize the genocides as carried out by patriotic heroes, Indigenous children must trace their family memories in exogenous photographic albums. For this reason, each photograph in *Fuegia* can be exhibited as a piece of history that disturbs memory and distances full recognition with an ominous past:

> When they had not been born, the demons of the night would suddenly come to the *kauwi* [Patagonian Indigenous tent] and beat the women. They screamed like savages, caused a lot of panic, and looted the camps. But unlike the breeders, they never killed anyone. She showed a crumpled photo: three men wearing masks, their entire bodies painted. In the center was her husband. "I got it from the museum professor," she told her children. "He took it out without anyone seeing him." After that she didn't say a word. She thought of the demons coming from the shadows to keep them terrified so that the day would not return to earth when women would rule again. Although the photo put an end to all that, her mother wept for the years she had lived deceived. But how to recognize one's own husband among a group of demons?

> Cuando ellos no habían nacido, los demonios de la noche llegaban de pronto a los *kauwi* y apaleaban a las mujeres. Gritaban como salvajes, provocaban mucho pánico y saqueaban los campamentos. Pero a diferencia de los criadores, nunca mataban a nadie. Ella mostró una foto arrugada: tres hombres poniéndose las caretas, con todo el cuerpo pintado. Al centro estaba su esposo. "Me la dio el profesor del museo," reveló ella a sus hijos. "La sacó sin que nadie lo viera." Después ya no dijo palabra. Pensaba en los demonios que llegaban desde las sombras para mantenerlas aterradas y que no volviera el día en la tierra en que mandaran otra vez las mujeres. Aunque la foto acabó con todo eso, su madre se lamentaba por los años que había vivido engañada. ¿Pero cómo reconocer al propio marido entre un grupo de demonios? (Belgrano Rawson 78)

The photograph represents the dividing line between documentary and personal history. Within it, the museum professor harbors an ominous return for the woman: "demons emerging from the shadows." This division raises the question of which historical gaze endures in the images. We can conclude that the album is the visual device that ethnographically segments the survival of the novel's protagonists. If we consider them through the lens of the opening photograph, the narratives question the status of the child's voice as Giorgio Agamben conceived it in *Infancy and History* (1979), where children inhabit the story of playability (58). In *Fuegia*, Belgrano Rawson narrates how the settlers force Indigenous children to tell the story of the massacre, not to play but to survive. The children's task, from this moment on, is to archive the memory of a devastated nation thanks to the shreds of history that museums keep in the form of photographs, taking from the shelves of oblivion those stories of Indigenous demons and shifting the signifiers toward a more accurate historical truth.

Bestiary: *La jaula de los onas*

La jaula de los onas by the Argentine writer Carlos Gamerro is set in Paris in 1889. The city is preparing for the great Universal Exposition, and opulent Argentina participates with a lavish pavilion in the French style, unaware that in the vicinity, a Belgian adventurer plans to exhibit in a cage, like wild animals, a group of Ona Indians captured in Tierra del Fuego and present them as "cannibals from Patagonia." One of the Onas, the young Kalapakte, flees the exhibition, dazzled by the Parisian lights and, wandering in the shadow of the newly erected Eiffel Tower, meets the young anarchist Karl, who participated in the construction of the megawork. Karl and Kalapakte, the former inspired by the ideas of the Franco-German revolutionaries and the latter by the stories presented by foreigners at congresses on Patagonian ethnography, embark on a return trip across Europe and Latin America, a return full of adventures and dangers that highlight the miseries and contradictions of a century that is ending and another that is just beginning.

Understood within the genre of documentary fiction, which Paula Klein defines as composed of "works that inscribe themselves on the boundary between literature and historical and archival research" [obras que se inscriben en el límite entre lo literario y la investigación histórica y archivística] (2), the novel presents five archival spheres and their counternarratives. These archival documents and narratives include the ethnographic photograph illustrating the book's cover; the social habitus of the upper and cosmopolitan classes of fin-de-siècle Argentina; the field of anthropology and social ethnography; the nineteenth-century anarchist archive with its struggles and its failures; and the language of the popular and migrant masses in the Southern Cone at the beginning of the twentieth century.

The novel's cover photograph shows the Selk'nam in front of the Eiffel Tower during the Universal Exposition. In the acknowledgments section, Gamerro traces the novel's inception to his encounter with the story of the captive Selk'nam through José María Borrero's *La Patagonia trágica* (1928), which recounts how they were kidnapped by the human trafficker Maurice Maître. The photograph, therefore, represents what Donna Haraway points out regarding the emergence of dioramas, the main form of exhibition at the Museum of Natural History in New York, which appeared when the restoration of the historical past is indicated by molecular genetic theories: "Restoration of origin, the task of genetic hygiene" (238). In this sense, pavilions begin to be understood as "realistic art," "windows to nature," or teleological national histories ["arte realista"; "ventanas a la naturaleza"] (Díaz Ángel 26).

The rupture of the dioramic transparency of the photograph occurs thanks to the appearance of an exogenous element in the Fuegian context: a white man dressed in a tailcoat who controls the Patagonians. Thus, we understand that the cover photograph, taken from the historical collection of human zoos, shows the nine Selk'nam who were forcibly taken to Paris for the Universal Exposition of 1889 to be exhibited in cages and pavilions (eleven were initially kidnapped, but two of them died on the outward journey). The cover image, fundamental to the narrative, exhibits a storytelling mechanism that places the novel alongside contemporary archival fictions: the use of photography to control and reproduce Indigenous and Latin American identity from the mid-nineteenth to the early twentieth century in the Western world.[6]

The novel's third chapter corresponds to an interview made in the early 1960s with the last survivor of the group, Rosa. Mediated by the voice of Felisa, a Selk'nam woman who translates, Rosa describes a photograph of the exposition's historical archive:

> See, here he is [the Frenchman in tailcoat]. That little dog they also brought. This is your brother Kalapakte, the one who stayed in Paris, right? . . . And here she is, this little girl here. You were *úlichen,* pretty, now you are *yíppen,* ugly like me, but older, Shemiken was her name, the Sisters called her Rosa later, in the mission, Rosa of Paris, but she likes her name better: Rosa Shemiken.
>
> ¿Ves? Acá está él [el francés en frac]. El perrito ese que también llevaron. Este es tu hermano Kalapakte, el que se quedó en París, ¿no? . . . Y acá está ella, esta niñita acá. Eras *úlichen*, bonita, ahora estás *yíppen*, fea como yo, pero más vieja, Shemiken se llamaba, fueron las Hermanas que le pusieron Rosa después, en la misión, Rosa de París, pero a ella le gusta más su nombre: Rosa Shemiken. (Gamerro 108)

For Rosa, the image contains that which was held captive. It is thus a witness of dispossession. Macarena Areco has elegantly argued that the portrait condenses Walter Benjamin's assertion about the documents of culture and barbarism: "There is no document of civilization that is not at the same time a document of barbarism" (Benjamin, "Thesis on the Philosophy of History," in *Illuminations*, 256; Areco 116–117). As Gamerro shows in the novel, modern societies capable of magnanimous monuments are simultaneously perpetrators of human miseries like caging and displaying Indigenous people. The nineteenth-century extractive reason is not fully satisfied with exploiting animals, plants, and minerals. It must also take possession of human production through a representational perspective, linking technological achievements with variety and freak shows (Arboleda 52).

In the novel, anthropology is another technology, in this case pedagogical, used by the French colonists to extract capital from Patagonian Indigenous bodies. In the chapter "A Letter from Franz Boas," dated July 14, 1893, the story is told of the German correspondent of Jewish origin in charge of the anthropology pavilion at the Chicago World's Fair, "the first in the world to have one organized according to strictly scientific principles" [la primera del mundo en contar con uno organizado según principios estrictamente científicos] (219). Thanks to the letter, we know that Boas writes constantly to his relatives, highlighting a project that aims to renew the discipline, creating "a modern and truly scientific anthropology" [una antropología moderna y verdaderamente científica] (227). Boas is suspicious of other anthropologists interested in the Fuegians, such as Roland Bonaparte, whom he describes as "one of those unhampered dilettantes who pretends to be a scientist and who, after taking his measurements in the most rigorous way and making his exact calculations and consulting his exhaustive tables, decided that [Kalapakte] was a polar Eskimo and sent them [Fuegians] north to Greenland" [uno de esos desahogados diletantes que pretende dársela de científico y que tras realizar sus mediciones del modo más riguroso y hacer sus cálculos exactos y consultar sus tablas exhaustivas decidió que (Kalapakte) era un esquimal polar y los mandó (a los fueguinos) al norte a Groenlandia] (223). Boas decides to undertake a photographic campaign to trace the true genealogy of the Selk'nam. In Chicago, Kalapakte observes one of these photographs from the Paris archive, on the back of which is written "cannibals of Patagonia captured by the Franco-Belgian explorer Maurice Maître" [Antropófagos de la Patagonia capturados por el explorador franco-belga Maurice Maître] (225). He recognizes himself in the photograph and sets out on the return journey to Patagonia.

Here again the Fuegians' memory work occurs when reading an exogenous archive made by European settlers. In the novel, Rosa is the survivor responsible for her Indigenous community's social reconstruction. She recounts how recognizing family traits and genealogical trees in the photographs of the Selk'nam archive taken and collected by anthropologists such as Martín Gusinde or in the European archive of Maître offers the only possibility of generating blood and community relationships for those close to the survivors. Contrary to the approaches to the family gaze typical of photographic albums that cut out the experience in favor of a private archive cared for by the descendants, Gamerro proposes a constellation of portraits that, as we intuit from the genocidal narrative, find no correspondence with the faces that accompany the story of Kalapakte and Rosa.

The Selk'nam, unable to access an archive of their own, must observe their familial affections through the foreign gaze. Kuhn points out that in narrating family photographs, "what I am telling you—'my own story'—about this image is itself changeable. In each re-enactment, each re-staging of this family drama, details are added and dropped, the story flashes out, new connections are made, emotional tones—puzzlement, anger, sadness—fluctuate" (17). Rosa, however, narrates her story, which, paradoxically, does not correspond to her because the one who adds and eliminates details about the album is not the consanguineous family member but the exogenous observer who administers the lineage as he pleases. However, this narration displaces the photographs from the representation of a human bestiary to bring them closer to an oral memory that reproduces itself in the voice of its survivors.

Catalog: *El museo de la bruma*

The novel-artifact *El museo de la bruma,* written by the Chilean author Galo Ghigliotto, builds a historic collection of objects and stories about large European landowners in Patagonia. In the manner of a museum of the imaginary, the narrative voice—in this case, the curator of the exhibition—invents the existence of a museum built in Tierra del Fuego after the end of World War II, extinguished in 2014 by a "mysterious fire" [misterioso incendio] that consumed both the facilities and the permanent collection (Gabrieloni 4; Ghigliotto 9). The curator of the narrative exhibition, obsessed with documents that prove the existence of the Museo de la Bruma, decides to undertake the task of an antiquarian and collector; he travels around Punta Arenas to find clues about the pavilions of the cultural center. The undertaking, at first impossible, comes to fruition once the curator finds "the only printed catalog of the museum" [el único catálogo impreso del museo] (9). With the catalog in hand "and the pieces that [he] was able to recover, [he has] put together this partial exhibition, perhaps phantasmagorical, that pretends to present the *aura* of the original exhibition of the Museo de la Bruma" [y las piezas que sí (pudo) recuperar (ha) montado esta exhibición parcial, acaso fantasmagórica, que pretende presentar el *aura* de la muestra original del Museo de la Bruma] (9). In structural terms and as detailed at the beginning of the text, the exhibition is divided into three biographical pavilions dedicated to tutelary figures of Patagonia: Julius Popper, an adventurer in search of gold; *Standartenführer* Walter Rauff, a Nazi officer who lived in Patagonia; and the British writer Bruce Chatwin alongside Mexican writer Alain-Paul Mallard, two explorers, mythmakers, and world travelers.

The novel presents itself as an investigative exercise comprising different types of documents. Among the archives are epistles such as that of the Selk'nam child (Item No. 9); paintings by Paul Gauguin (Item No. 224); photographs including the abhorrent Item No. 98, "A Fuegian Athlete" (1887), and of Selk'nam children shearing sheep; testimonies of various Patagonian citizens; CIA memoranda on the probable advice given by Walter Rauff to the DINA (the Chilean secret police during the Pinochet dictatorship) at Dawson Island (Item No. 11); invoices for the purchase of Indigenous bones by the British Museum; food records of Julius Popper's expeditions; records of Indigenous deaths at Mission San Rafael; gold teeth extracted from decomposing bodies; newspaper clippings; juridical summaries; references to books such as the compilation by journalist Virginia Vidal chronicling how the famous Patagonian writer Francisco Coloane attended Rauff's funeral and bid him farewell with Nazi greetings; and a letter from President Salvador Allende to Simon Wiesenthal written in 1972 (Item No. 31) regarding the Rauff case and the failed attempt to extradite him to Germany to be judged.

Given its affinity with the space of the archive and the problematics of the museum, *El museo de la bruma* has been read as belonging to the postautonomous writing practices that, according to Josefina Ludmer, cross "the borders of literature" [las fronteras de la literature] (56). It has also been studied through Florencia Garramuño's concept of contemporary unspecific art practices that do not easily fit into the traditional partition between disciplines (15). As the leading proponent of this reading, Macarena Areco places Ghigliotto's work within that of a more extensive group of Chilean writers—Cynthia Rimsky, Nona Fernández, Matías Celedón, Juan Pablo Sutherland, and Alejandra Costamagna—that can be labeled under the concept of "archival fictions" [ficción de archivo] (111).[7] What brings these literary works together is the presence of archival elements, many of them institutional, cataloged by subjects who experience generational traumas.

What is unprecedented in Ghigliotto's work is that together with an absent or depersonalized narrative voice—it is the assembly of works that delivers an apparent subjectivity—we observe an archival work that escapes the norms of the testimonial genre, since the discourses and objects lose the apparent transparency of institutional accounts and are subject to suspicion (Singer González

75). Deviating from Popper's methodology regarding the cataloging of the Patagonian Fuegian people, in which the optical regime of the photographic album forecloses Indigenous identity, *El museo de la bruma* recovers legends about unknown Fuegian peoples of whom we have no photographic records (Andermann 58). At the same time, it imagines new narratives of guilt regarding genocides in southern Chile. If, as Jens Andermann suggests, the taxidermic transparency of the photographs of Fuegians is one of the constitutive principles of their existence (61), Ghigliotto gives them the opacity necessary to generate new narratives and kinships.

On the one hand, the narrative voice presents the result of the search for the museum's collection; on the other, the result is the articulation of a series of elements gathered under the title "El Museo de la Bruma" (The Museum of Fog). At the intersection of both fictional operations, the narrative voice acts as a curator of various documents. In this perspective, it is essential to inquire about the link that unites the series of archives maintained by the Museum of Fog and the institution's collecting device or, in other words, to ask the question of taxonomic relevance: the intersection between the fiction of such institution and the institution within the fiction itself. Regarding this general operation, we can mention Item No. 134, a jar with genitalia that was kept in the Salesian school of Punta Arenas: "For a long time it was believed that these testicles belonged to a bull bovine; however, later studies revealed that they are human organs, specifically from an adult male of large stature and about thirty years of age" [Por mucho tiempo se creyó que estos testículos pertenecían a un vacuno semental; sin embargo, posteriores estudios dieron a conocer que se trata de órganos humanos, específicamente de un hombre adulto de gran talla y unos treinta años de edad] (247). The institution's information is not definitive but comprises various clues: "A clue is given by Salesian Father Alberto de Agostini" [Una pista la entrega el padre salesiano Alberto de Agostini] (247). We learn that during the process of colonial expansion in Patagonia, colonists paid for the genitals and ears of murdered natives. The text does not provide a conclusion; it only indicates the existence of both documents, the jar and the clue.

If no progress is made toward conclusions, it is because of the book's structure. At the beginning of the text the narrator explains, "In terms of the museum's order, in any case, we could not elucidate a logical coherence but rather esoteric features" [En términos de orden, en todo caso, no pudimos dilucidar una coherencia lógica sino más bien rasgos esotéricos] (9). Instead of claiming a finished—even causal—logic, the novel displays the writing of an unwritten story. Thus, the artifact-like character of the novel reflects its thought operation;

the voice wanders among documents in an unwritten account, highlighting the existence of such an account and ensuring its permanence for posterity. The novel does not pretend to clarify a past but to work with the consistency of the fog. The museum's esoteric features are in the catalog's gaps and the fine connection of horrors that emerges between objects and clues. The unspoken narrative of the museum's curatorial zeal encompasses the modern history of Patagonian horrors.

Toward a Conclusion

Belgrano Rawson's photographic album, Gamerro's bestiary, and Ghigliotto's catalog inquire into the disparity of an event; the appearance of Tierra del Fuego in the universal exhibitions, within the repertoire of goods exhibited on a global scale, coincided with the time of the Indigenous Patagonians' disappearance, that is, their genocide. By adopting documents as writing platforms, the works we studied provide at least two problems concerning the relationship between exhibition and extermination. On the one hand, by considering the Universal Exposition as the presentation of caged humans in a museum key, these novels highlight the mortuary status of the exhibit space; the individuals on display do not flaunt the singularity of their lives but expose the expiration of their collectives. They are archaic specimens brought to life for this spectacular occasion. On the other hand, the Indigenous Selk'nam people displayed in the universal exhibitions—and whose stories are recovered by these contemporary writers—relate to the past from the double spatial matrix in which they were situated. In the killing zone that was Patagonia, the history of the Selk'nam is foggy, but their inclusion within global history (as a universal narrative) is pristine and transparent. In both cases, the Fuegian is only given to being seen as a distant past.

"World exhibitions glorify the exchange value of the commodity. They create a framework in which its use value recedes into the background. They open a phantasmagoria which a person enters in order to be distracted," writes Walter Benjamin in *The Arcades Project* (7). We cannot be indifferent to the quotation when we return to the living specimens displayed in the universal exhibitions, to their exalted exchange value and their receding use value. Even in this exhibition artifice, the Fuegians were cast as a phantasmagoria of an inhospitable region of the world, now visible and evident to the metropolis. To be included in the global story, the Fuegian people were stripped of themselves and thrown into the past. As living specimens, the Fuegians were spatially close to the distracted passersby but were temporally relegated to the distant past.

Authors' Note

This article was originally published as "El museo de la modernidad: archivo fotográfico, colección patrimonial y genocidio Selk'nam en Carlos Gamerro, Eduardo Belgrano y Galo Ghigliotto," by Ignacio Pastén López and Ignacio Veraguas Caripan, in *Cuadernos LIRICO* 26 (2024).

Notes

1 Colin Rowe and Robert Slutzky differentiate between literal and phenomenal transparency. See their chapter "Transparency: Literal and Phenomenal," 21–55.

2 In 1873 the so-called Exposición del Coloniaje, organized by the frantic mayor of Santiago, Chile, Benjamín Vicuña Mackenna, took place. The aim of the exhibition, which housed more than 600 pieces, was to contrast the city's progress with the country's colonial vestiges. Among the pieces in the exhibition was José Esti, a native of Tierra del Fuego recently imprisoned by the governor of the colony of Magallanes. Regarding this "living specimen," the newspaper *El Independiente* on September 17, 1873, published the following: "It is hard to believe, but you must believe it or bust, the Patagonians, with their disgusting costumes and their repellent figures, are monopolizing the attention of the curious. They are taken to the exhibition . . . and even into some respectable houses" [Es para no creerlo pero hai que creerlo o reventar, los patagones, con sus asquerosos trajes i sus repelentes figuras, están siendo los leones de curiosos i curiosas. Se les lleva a la esposicion . . . i hasta se les introduce en algunas casas respetables] (quoted in Acuña 10). Unless otherwise stated, all translations are the translator's own.

3 The structure of the Chilean pavilion was the responsibility of Plenipotentiary Minister Carlos Antúnez during the government of José Manuel Balmaceda. Given the construction costs, a removable and dismountable iron, steel, and zinc structure was commissioned to be reused once the Universal Exposition was over. The reconstructed building currently houses the Artequin Museum. It should be noted that the construction of glass and iron as a symbol of civilization and power for the oligarchic class is also reflected in the Quinta Normal Greenhouse, purchased by the state from a private individual and installed in 1890.

4 The German priest and anthropologist Martín Gusinde (1886–1969), who sought to investigate the different Indigenous peoples to test and prove that the belief in a single, monotheistic God was connatural to human beings, is noteworthy in the Chilean case. The priest himself points out that the documentary appearance of the Indigenous people coincides with their disappearance as a human group. In 1937 Gusinde declared, "However much one may pity it, the peculiar little people of the Yámana have disappeared forever" [Por más que se lo compadezca, el peculiar pueblito de los yámana ha desaparecido para siempre] (quoted in Pavez Ojeda 261). See also Palma Behnkse.

5 Hirsch proposes the "family gaze" as the organizing vision of family memories (xii). In contrast, Kuhn proposes the "family secret" as the device that watches over the memories of the descendants (15).

6 In an interview for Argentine newspaper *Página 12,* Gamerro refers to the importance of this image: "The idea of those Indigenous people from Tierra del Fuego, those in the photo that opens the book, coming from a people of hunters and gatherers who had had practically no contact with whites until that moment, suddenly finding themselves in the capital of the nineteenth century, at the foot of the Eiffel Tower, a construction that wanted to represent the summit of human progress. That conjunction, I thought, was too good to be left aside. Something had to be done. It remained there, gestating for thirty years" [La idea de esos indígenas de Tierra del Fuego, los de la foto que abre el libro, provenientes de un pueblo de cazadores y recolectores que no habían tenido prácticamente contacto con los blancos hasta ese momento, que, de golpe, se encuentran en la capital del siglo XIX, al pie de la Torre Eiffel, construcción que quería representar la cumbre del progreso humano. Esa conjunción, pensé, era demasiado buena como para dejarla de lado. Había que hacer algo. Quedó ahí, gestándose durante treinta años] ("La novela de Carlos Gamerro").

7 Areco describes the main features of contemporary archival fictions as being the importance of family and photography, the inquiry into genealogical injustices and pains, the juxtaposition of history and fiction, and the trope of the journey, whether in the form of migration or exile (112).

Works Cited

Acuña Fariña, Constanza. "La fortuna crítica del arte colonial en Chile: Entre la avanzada del progreso, la academia y la sobrevivencia del pasado." In *Perspectivas sobre el coloniaje,* edited by Constanza Acuña Fariña, 7–19. Ediciones Universidad Alberto Hurtado, 2013.

Agamben, Giorgio. *Infancy and History: On the Destruction of Experience.* Translated by Liz Heron. Verso, 2007.

Andermann, Jens. *The Optic of the State: Visuality and Power in Argentina and Brazil.* University of Pittsburgh Press, 2007.

Arboleda, Martín. *Planetary Mine: Territories of Extraction under Late Capitalism.* Verso, 2020.

Areco, Macarena, "Ficción de archivo: viaje, enigma, memoria e historia." In *Ficción—No ficción en América Latina,* edited by Macarena Areco, Fernando Moreno, and Cécile Quintana, 111–119. Éditions des archives contemporaines, 2024.

Artiles Burgos, María de la O., and Francisco Javier Boned Purkiss. "Arquitectura moderna y transparencias: materialidad y virtualidad." *Artnodes: Revista de Arte, Ciencia y Tecnología,* no. 15 (2015): 47–60.

Barthes, Roland. *La Torre Eiffel: Textos sobre la imagen.* Translated by Enrique Folch González. Paidós, 2001.

Bergot, Solène, and María José Correa. "Chile y la escenificación de su modernidad: Ciencias y técnicas en las Exposiciones Universales Nacionales (1869–1888)." In *Ciencia y espectáculo: Circulación de saberes científicos en América Latina, siglos XIX y XX,*

edited by María José Correa, Andrea Kottow, and Silvana Vetö, 47–69. Ocho Libros, 2016.
Belgrano Rawson, Eduardo. *Fuegia.* Seix Barral, 1991.
Benjamin, Walter. *The Arcades Project.* Translated by Howard Eiland and Kevin MacLaughlin. Harvard University Press, 1999.
Benjamin, Walter. *Illuminations.* Translated by Harry Zohn. Schocken, 2007.
Comeron, Octavi. *Arte y postfordismo: Notas desde la fábrica transparente.* Trama, 2007.
Díaz Ángel, Sebastián. "Mapas y dioramas: Elementos para repensar la construcción cartográfica de la naturaleza y la naturaleza de los mapas." *Terra Brasilis* 4 (2015): 17–31.
Foster, Hal. *The Art-Architecture Complex.* Verso, 2013.
Gabrieloni, Ana Lía. "Imprecisiones a través de la bruma de un museo austral o el sublime paisaje de la memoria de los tiempos." *Actas de las I Jornadas Nacionales Espacios, afectos y nuevas formas de lo cotidiano,* 1–7. Conference proceedings, Facultad de Filosofía y Letras, Universidad de Buenos Aires, December 16–17, 2020.
Gamerro, Carlos. *La jaula de los onas.* Alfaguara, 2021.
Gamerro, Carlos. "*La jaula de los onas,* la novela de Carlos Gamerro que reescribe el discurso de Civilización o Barbarie." Interview, *Página 12,* June 27, 2021. https://www.pagina12.com.ar/350413-la-jaula-de-los-onas-la-novela-de-carlos-gamerro-que-reescri.
Garramuño, Florencia. *Mundos en común: ensayos sobre la inespecificidad en el arte.* Fondo de Cultura Económica, 2015.
Ghigliotto, Galo. *El museo de la bruma.* Santiago, Chile: Laurel Libros, 2019.
Haraway, Donna. "Teddy Bear Patriarchy: Taxidermy in the Garden of Eden, New York City, 1908–1936." In *Cultures of United States Imperialism,* edited by Amy Kaplan and Donald E. Pease, 237–291. Duke University Press, 1993.
Hirsch, Marianne. *Family Frames: Photography, Narrative, and Postmemory.* Harvard University Press, 1997.
Jostic, Sonia. "El relato exasperado: iconización de la palabra y narrativización de la imagen." *Alba de América* 29, no. 55–56 (2010): 277–293.
Kay, Ronald. *Del espacio de acá: Señales para una mirada latinoamericana.* Metales Pesados, 1980.
Klein, Paula. "Poéticas del archivo: el 'giro documental' en la narrativa rioplatense reciente." *Cuadernos LIRICO* 20 (2019): 1–13.
Kuhn, Anette. *Family Secrets: Acts of Memory and Imagination.* Verso, 1995.
Ludmer, Josefina. *Aquí América Latina: Una especulación.* Eterna Cadencia, 2010.
Lyotard, Jean-François. *Libidinal Economy.* Translated by Iain Hamilton Grant. Indiana University Press, 1993.
Palma Behnkse, Marisol. *Fotografías de Martín Gusinde en Tierra del Fuego (1919–1924): La imagen material y receptive.* Ediciones Universidad Alberto Hurtado, 2013.
Pavez Ojeda, Jorge. *Laboratorios etnográficos: Los archivos de la antropología en Chile (1880–1980).* Ediciones Universidad Alberto Hurtado, 2015.
Peña, María Antonia, and Rafael Zurita. "The Peruvian Native and the Conception of Liberal Citizenship in the Latin American Context." In *Enemies Within: Cultural Hierarchies and Liberal Political Models in the Hispanic World,* edited by María Sierra, 7–40. Cambridge Scholars, 2015.

Perkowska, Magdalena. *Pliegues visuales: Narrativa y fotografía en la novela latinoamericana contemporánea.* Iberoamericana, 2013.

Pratt, Mary Louise. *Imperial Eyes: Travel Writing and Transculturation.* Routledge, 1992.

Rowe, Colin, and Robert Slutzky. *Transparency.* Birkhäuser, 1997.

Singer González, Deborah. "El testimonio de Rigoberta Menchú: Estrategias discursivas de una subjetividad fronteriza." *Revista Latinoamericana de Derechos Humanos* 23 (2012): 73–88.

Sloterdijk, Peter. *En el mundo interior del capital: Para una teoría filosófica de la globalización.* Translated by Isidoro Reguera. Siruela, 2006.

Tello, Andrés Maximiliano. *Anarchivismo: Tecnologías políticas del archivo.* La Cebra, 2018.

Uslenghi, Alejandra. *Latin America at Fin-de-Siècle Universal Exhibitions: Modern Cultures of Visuality.* Palgrave Macmillan, 2016.

von Tschilschke, Christian, and Dagmar Schmelzer. "Docuficción: Un fenómeno limítrofe se aproxima al centro." In *Docuficción: Enlaces entre ficción y no-ficción en la cultura española actual,* edited by Christian von Tschilschke and Dagmar Schmelzer, 11–34. Iberoamericana, 2010.

2

Árbol Ramón, the Axolotl/Ajolote, and LEGOs

Mexico City's Papalote Children's Museum

Emily Hind

My analysis centers on a children's museum in Mexico City, Papalote Museo del Niño, a private institution that tackles some themes previously assigned to natural history museums. In the study I consider two naturalist exhibits at Papalote: one employee-led tour about the food chain that has children and their adult companions clamber through and around a multistory human-made tree, Árbol Ramón, and one self-guided set of activities devoted to an amphibian species under threat of extinction, the axolotl or ajolote. These two competent biology lessons operate in conjunction with the LEGO activities found on the same ground-floor exhibition space. A LEGO city winds up structuring what I understand of the biology lessons: the museum comes to seem impossible to exit in any definitive sense, while the activities reassure visitors that a safe space might exist somewhere, if parents and children create it. This commanded creativity and vigilance, within the museum guidelines at least, repeat structures that already rescind possibilities of a safe nest. I wrap up the essay with a review of museum ponderings in Spanish-language texts on the axolotl by deceased writers Julio Cortázar and Salvador Elizondo and by living writers Rafael Lemus and Andrés Cota Hiriart. The tradition of the natural history museum, as legible through Papalote's LEGO city, indicates that the complexity of sustainable ecologies vastly exceeds the type of solutions that we can build from the items, such as plastic, that cause the problems in the first place.

There are no LEGOs at the public Natural History Museum of Mexico City, which is not especially aimed at young children. The older intended audience does not make it easier to run the museum, however. Even among its own experts, this institution elicits a frustrated defense. Despite the target museum audience's ignorance of the natural world and the looming threats of the Anthropocene, funding is meager: "One comes to work every day in a museum devoted to nature, without resources, whose scientific and exhibition guidelines date to 1964" [uno llega todos los días a trabajar en un museo de la naturaleza, sin recursos, cuyo guión científico y museografía datan del año de 1964] (Vázquez Martín 112). Eduardo Vázquez Martín wrote those words in an essay published in 2011 in which he grumbles about the ongoing remodeling of the Natural History Museum that had begun in 2006. When I visited the museum some seventeen years later, in early March 2023, most of the one-story dome-shaped buildings the museum comprises were still closed for refurbishing. I did admire the exhibits, which were smartly organized and quite detailed, as they discussed climate change and extinction events.

Papalote more successfully engages small children than the Natural History Museum does. The privately funded, nonprofit children's museum benefits from corporate philanthropy and much higher admission costs, and it features well-staffed interactive lessons. A reflexive explanation for the comparatively underfunded Natural History Museum might proffer a glib reference to neoliberalism, a reflex that I view as shortchanging the vast sweep of the Anthropocene. I neither advance nor withdraw the explanation of neoliberalism when I state that a wall display of donors to Papalote includes the American Museum of Natural History. I do want to keep in mind Carolyn Fornoff's keen analysis of the imbrication of cultural institutions with the problems of the Anthropocene such as extractivism; critique of this relationship warns against complacency with both "the infrastructures of art, and how art in turn has contributed to the naturalization of extractivist logics" ("Reflexive" 63). Fornoff's considerations on art institutions apply here; natural history displays make use of art technique, as Dolly Jørgensen's pondering of taxidermy-mediated encounters with the extinct bluebuck reminds us ("Portraits of Extinction"). The infeasibility of withdrawing from this imbrication with the problem, whether termed "neoliberalism," "extractivism," or, writ large, "the Anthropocene," requires that we extend grace toward museum visitors and directors. Rather than condescension toward Mexican institutions, my attitude, however exasperated, means to turn toward Fornoff's *Subjunctive Aesthetics* with its sidestepping of certainty in favor of possibility.

I see no reason to view underfunding of the Museum of Natural History as a foregone conclusion, even under neoliberal arrangement. Public projects today can receive lavish subsidies, as evinced by the comparative largesse that the Mexican government has shunted toward mega-endeavors undertaken during Andrés Manuel López Obrador's six years in office, 2018–2024, such as building the pro-pollution oil refinery Olmeca and the Maya Train. Note that the president, AMLO, distracts us from the stupidity of his ecologically dangerous projects by muttering bitterly about the "neoliberal" as if this term sufficed now as an archenemy, now as a totalizing explanation. My approach here means to reject the feedback loop between overgeneralizing academic habit and that vacuous presidential rhetoric, and I mean to heed Elizabeth Anker's concerns regarding theory mired in paradox. *On Paradox* has Anker articulate the academic love of contradiction as "both problem and solution" (10), both "riddle and key" (137) that threatens to paralyze us in a "haze of indecision" (23) and that "does nothing to delimit or prescribe what should come along to replace whatever principles it has abolished or cleared away" (87). To judge from AMLO's discourse, proposing the problem for the solution can coincide with also blaming others who do this as if one were not simultaneously committing the same folly. By contrast, I take seriously Anker's warning that the fashion of theoretical paradox leads us into a trap. Of course, my thoughts on Papalote explore paradox at length, but in no instance do I mean for us to stay there as some kind of final enlightenment. We need change.

With this framework in mind, let me begin again, with more precise accounting for the price of entry to these museums. The Natural History Museum in June 2025 charged 38 Mexican pesos for an adult, about US $2, and children paid half that. In June 2025, a ticket to Papalote that included access to a short film cost 252 pesos. This price was listed as an automatically discounted fee from the actual price of 280 pesos. The required adult companion ticket doubled the minimum price for one child to 504 pesos, some US $26.50. The lowest ticket price, without access to the film, cost 220 pesos per ticket, bringing the minimum total for one child to 440 pesos or about US $23. I have rechecked those data periodically and sometimes adjusted the numbers upward. Despite the price discrepancy, many Mexicans have recommended that I visit Papalote, while no one has ever said I should check out the Natural History Museum. The latter houses a few displays that caught my son's interest, such as a touchscreen video presentation on climate patterns, but at nearly six years of age, he was unable to read the complex explanations on the screen and the accompanying mounted displays. There is simply no way I could have entertained him for hours there. Only Papalote has operated consistently on my son's intellectual level, as per the museum motto, "touch, play, and learn" (Servitje de Lerdo de Tejada 121). Probably more visitors patronize Papalote as a result.

In the 2011 essay, Vázquez Martín calculates that annually, "almost four hundred thousand visitors" [casi cuatrocientos mil visitantes] visit the Natural History Museum run by the "small and heroic team of staff members" [pequeño y heroico equipo de trabajo] (112). The same collection of essays includes a piece on Papalote that claims an annual total of "almost two million visitors"

(Servitje de Lerdo de Tejada 122). That last number may be hyperbole, though it accurately identifies the stronger reputational luster of Papalote. If two million guests did arrive at Papalote by 2011, the numbers have dwindled. Certainly, the annual report on the Papalote website for states that by 2023 more than 23 million people had "experienced the magic of Papalote Museo del Niño" [experimentado la magia de Papalote Museo del Niño] (Papalote, *Informe anual 2023*, iv). However, in 2023 this same report acknowledges a total of 458,000 people who visited the museum that year (30). That number reflects a significant decrease in attendance compared to 2014, the date of Papalote's earliest annual report available on the same museum website, when Papalote welcomed a total of 699,386 visitors (Papalote, *Informe anual 2014*, n.p.). Some guests access the grounds for free, and the reports cast around for appropriate euphemisms to name these sponsored guests. The 2023 document mentions service to "vulnerable communities" [comunidades vulnerables] (33), while the annual report from 2014 refers to welcoming visitors hailing from the "basic level" [nivel básico] (n.p.). The struggle to name the poor already hints at the considerable obstacles to inclusion.

I never visited Papalote prior to its extensive remodeling in 2016, which updated the installations originally opened in 1993. The children's museum as I know it has always featured the contradictory renovation of 2016, with more efficient cooling and wastewater treatment systems, along with green areas, exhibition halls, and a parking garage (Adrià 36). Note the tension there: a parking garage increases the very consumption that the conservation-oriented efforts at Papalote claim to want to remedy. The parking garage anticipates the fact that Papalote does not unequivocally combat the problems of economic disparity, which is to say that it participates in the problems that it also sets children to solving. Politically committed LEGO play illustrates this gesture.

LEGO City and the Anthropocene: Child's Play?

LEGOs at Papalote reinforce, in theory, the lessons on environmentalism attached to the food chain and the axolotls; the LEGO play may signal the limits of the natural history museum. Permanent cue cards at Papalote, bolted to the countertop, assign children the task of imagining sustainable design regarding either water or transit through LEGOs. Perhaps few young visitors read the directions. Papalote has a large number of employees to guide the interactions and make the museum experience for Spanish-language speakers fundamentally different from the trip to the Natural History Museum. These employees, dressed in bright yellow vests and known in the youthful parlance of Papalote as *cuates* (buddies), facilitate group activities for children in ways that make the museum a bit more than the sum of its material parts. At least one employee is stationed at nearly every activity at the children's museum. Admittedly, I have visited Papalote four times with my son—and after the birth of my daughter in 2023, with an infant as well—not just because of the *cuates*. I lean on Papalote because of the autocentric design of Mexico City. Car culture makes it difficult to entertain a child safely in the capital. Papalote offers kids a full day of safe play, in compensation for the unfreedom of children in Mexico City caused by traffic-heavy streets and comparatively few parks, a problem that the museum parking garage only exacerbates even as the interactive exhibits such as the LEGO activity warn against depending on driving cars.

Not to drum on about periodization, but please note that the inequities of car culture predate neoliberalism by a half century. The lessons in Papalote most explicitly center on ecology and sustainability—lessons that might comfortably appear in the contemporary natural history museum—and coincide with a resource-intensive complex in ways that speak to habits far beyond those of the past five decades. To understand this design, I turn to Jonathan Lee's book on LEGOs, in which he explains that by contrast to the expansive interior spaces of dollhouses, "most LEGO housing play privileges exteriors over interiors" (36). During my time in the glassed-in LEGO playroom of Papalote, I have admired the intricate, unchildlike LEGO city already constructed and protected in the center of the room behind plastic walls: a LEGO city. Unlike my fascination with dollhouses and certain children's book illustrations of cozy interiors, I have never imagined living among those LEGO models, however. In another passage, Lee articulates a convincing reason for that absent daydream. Taking inspiration from Elizabeth Grosz's *Architecture from the Outside,* Lee writes that LEGO play winds up in "a world of our own creation that has no room for us" (36). Papalote curtails the problems of a world that has no room for children, so as not to upset them, but also assigns them angles of the predicament to play at solving.

In thinking about the open design of Papalote and its heavy use of glass, I find it helpful to review Lee's summation of a particular playset, the "3-in-1 LEGO Creator Treehouse," which boasts an architecture of openings but no doors and thus creates continuity between interior domestic realms and exterior natural ones. As Lee observes in the case of the LEGO treehouse, the flow between indoor and exterior design has the effect of eliminating any "material distinction between architecture and nature" (36). In Papalote, the limits of the natural history museum seem to pressure if not outright collapse would-be distinctions between nature and human design, crushing the hope for a longed-for safety of the inside by presenting an endless series of exteriors, which nonetheless hint at a worse problem just beyond those seamless borders of the museum.

Possibly because of poor timing that has my son enter the room after the official start of the timed activity, he has always preferred to build his own LEGO designs, unrelated to the assignment explained in the cue card that starts out in all capital letters: "We are running out of water!" [¡Nos estamos quedando sin agua!]. The directions implore LEGO builders to devise a solution: "Imagine as a team a great solution to have more water and build it with bricks" [Imaginen en equipo una solución genial para tener más agua, y constrúyanla con bricks]. The unitalicized use of the English-language term "bricks" echoes an Anglophone concept of "maker" presented elsewhere in the museum. A list of ten principles stenciled on a wall ends with an unenumerated declaration: "I can be a maker and prove it" [¡Puedo ser un maker y demostrarlo!]. The term "maker" insinuates, for my ear at least, a kind of Silicon Valley entrepreneurship, an interpretation that finds support in the engineering ethos described in the first of the ten principles: "1. I enjoy learning by doing and solving challenges" [1. Me gusta aprender haciendo y resolviendo retos].

As the museum has evolved, another set of ten principles has appeared on the museum wall, though the old rules have remained posted too. These new ideals, expressed in no particular order thanks to the visually networked layout, suppress the term "maker" and conjure girl and boy Papalote Creators ["Una Creadora y un Creador Papalote"] who adhere to goals like fomenting creativity, education, and sustainability, along with promoting innovation: "Fomenta la creatividad; Fomenta la educación; Fomenta lo sustentable; Promueve la innovación." Adding concern for sustainability as a somewhat isolated concept may do little to refocus lingo associated with capitalist maker types who favor profitable innovation and creativity. More tellingly, the absence of technofixes to the problems stirred by technology already hints at a certain hollowness in the mandate to "Promote innovation." How a child might fix the water emergency never finds an answer anywhere that I saw in the museum and certainly seems to be absent from the model LEGO city. Still, the recognition of sustainability within the new set of values signals yet another shift toward the concerns associated with the natural history museum. These values, as I keep insisting, exist in tension with the sort of play that a children's museum might be assumed to champion. For an understanding of the escape valve that the museum carves out, I return to Lee's musings on LEGOs.

On the one hand, Lee explains, the design of LEGOs, as a product of the mid-twentieth century, reflects "a suburban ideal in which entire planned, cultivated, and regulated suburban communities function like fully protected domestic spaces" (43–44). On the other hand, Lee argues that companion LEGO toys aimed to alleviate the possibility of asphyxiation under this full protection because they offered the player "an explosion of mobility" through "a train, helicopter, tow truck, delivery truck, three cars, a scooter, a bicycle, a hot dog cart, and a handcart," which all counterbalanced "LEGO City's erosion of space" (52). The LEGO city protected behind the plastic shield in Papalote includes numerous vehicles, none of which addresses, at least not in any obvious fashion, the second assigned problem in the LEGO room instructions, again in all capital letters: "We need more and better transportation!" [¡Necesitamos más y mejor transporte!]. The directions call for children to invent a way to transport more people ("más personas") and end with the word "bricks." The model city provided seems to include the usual toys but no novel solutions. In fact, in the twenty-first century, the Danish-based LEGO company initially rejected a fan's request to add bike lanes; the company dismissed the request as a code violation because LEGO pieces must avoid political statements (Gordon and Steeman). Still, mobility matters in LEGO play, hence all the germane transportation toys.

The impossibility of getting into that model LEGO city behind Lucite walls at Papalote supports Lee's provocative points. The museum's efforts to address the disaster of the Anthropocene provide a maker's play space and repeatedly issue the request that children reengineer this design. Children are to develop solutions from the imagined outside, a safe exterior, vis-à-vis the core problems. In reality, transportation and water shortages, along with climate change, air pollution, and other terrible threats, spare no one; from inside the museum, we are already in the problem. At the zoomed-out level of the Anthropocene, there is no true exterior, only concentric rings of privilege that orbit inside the gravitational pull of the problem at varying degrees of intensity. The exteriors implied by LEGO design facilitate the hint in Papalote regarding the safety of the museum as a place of experimentation distanced from the immediate risks that children are to imagine resolving through engineering play. If children were to solve these problems, they might upend the very design of Papalote, which has much in common with the traditionally car-dependent architecture of LEGO city. To get at this parallel between Papalote and the Anthropocene accelerated in LEGO infrastructure, I point to the example of the food court.

(I know that you don't want to read about the food court, but allowing you to skip exactly where you will end up when you visit a museum for hours on end with kids betrays our would-be commitment to them.) Sustainable choices are not dominant among the franchise menus on the open-air patio of Papalote that include a Mexican sandwich (*torta*) stand, a sushi restaurant, Domino's Pizza, Burger King, and Kentucky Fried Chicken, if by sustainable we mean vegan. A mother, perhaps feeling the weight of the Anthropocene, might choose side dishes like rice at the sushi stand or french fries at Burger King, but other seemingly vegan items probably contain some sort of unsustainable animal product; beware of the milk in the mashed potatoes from KFC and the probable lard in the beans from the *torta* shop. The franchises propose a carbon-heavy, trash-producing meal that may quash hope of stepping outside LEGO city, any illusion of moving from the outdoors patio into a natural world, though we are already, to the vanishing degree possible, "out of doors." To be sure, I do not blame the museum administrators in particular for the squandering food-chain choices at the food court. I blame *Homo sapiens* and find support for this species-wide mistake in Yuval Noah Harari's planetary history titled *Sapiens,* in which he observes that long before the Industrial Revolution, humans collectively held the record for sending the most plant and animal species into oblivion: "We have the dubious distinction of being the deadliest species in the annals of biology" (89). We also have the dubious distinction of forgetting our lethal ways.

A helpful companion piece to understanding Harari's *Sapiens* is Jørgensen's writing on short human memories and extinction. In her book *Recovering Lost Species in the Modern Age,* Jørgensen points out that people tend to imagine, incorrectly, their childhood as a normative accounting baseline for species (135, 139). Rewilding efforts can thus be ignored entirely by new generations because the reintroduction of a species becomes for youngsters a mistaken experience of an unbroken timeline of habitation. This shaky grasp of natural history surely applies to the baseline experience of the section of Chapultepec Forest where both Papalote and the Natural History Museum are located. This domesticated space that has been stripped of (nonhuman animal) predators to humans must seem like true wilderness to the young in the megalopolis. Not to make a pun on the concept of rewilding, but the most feral edge of this space is the predatory layout for the poor, with its inaccessible or only intermittently accessible grounds for many economically disadvantaged would-be visitors. Just as there is no beans and rice menu item among the franchises, there isn't really any easy mode of sustainable transit to the buildings amid this seemingly open public park, echoing Lee's warning about exteriors in LEGO design that eliminate any "material distinction between architecture and nature" (36).

Ultimately, the LEGO city hands us a functional interpretation of the sprawling forest that in total, across all three sections of the park, "represents more than 50 percent of the city's total green area" [representa más del 50 por ciento de las áreas verdes totales urbanas] (Escotto 259). This green area strikes me as just more maker space, squeezing us out with excessive emphasis on exteriors. For readers unfamiliar with the nearby roads, I should explain that the low-traffic second section of Chapultepec Forest is bounded by congested roads decidedly unfriendly to pedestrians and cyclists. The north-south axis of the section abuts Constituyentes Avenue and the northwest-southeast axis, the Periférico beltway. Even the part of Paseo de la Reforma that bounds Chapultepec in this area is hostile to pedestrians and cyclists.

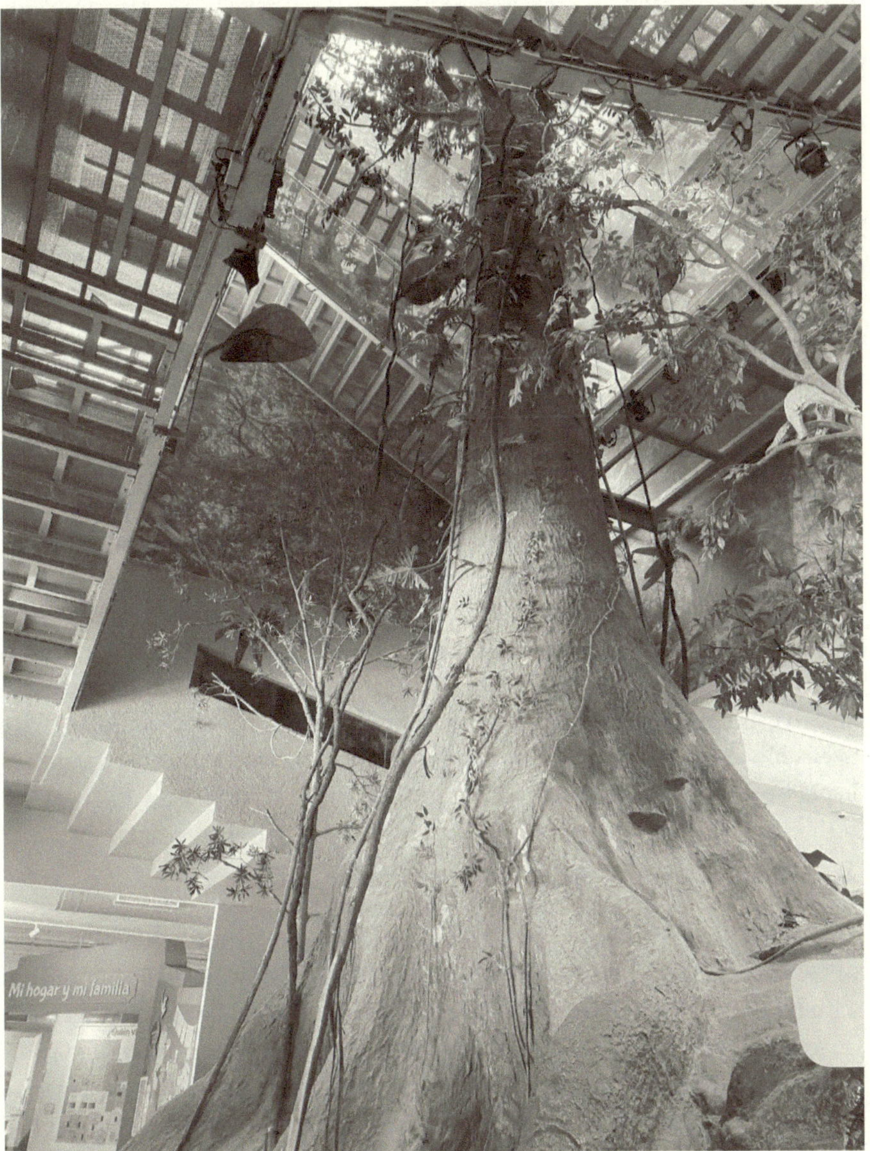

Figure 2.1. Replica of the *ramón* tree, *Brosimum alicastrum,* in Papalote Children's Museum. Photograph by the author.

Once inside the open air of the second section, cyclists and pedestrians can enjoy the largely empty, meandering paved roads; getting there is the problem, which makes even the forest another type of exterior. But once inside, there is still no true safe haven, no nest. As a result of this bizarrely isolated public space in the middle of the densely populated capital, the second section hosts only 26 percent of annual visitors, who generally arrive by automobile, indicating relatively elite status (Garduño Serrano 352).[1] There is no exit from the layers of design that swaddle Papalote and that limit the ideals of its exhibits. Perhaps fortunately for children's mental health, the museum, always surrounded by another layer, encourages the LEGO player to operate as if from the exterior of a severe planetary emergency—still able to consume as many hamburgers and Styrofoam cups as a personal budget allows. The domesticated space available in each of these layers of unending architecture and eroded nature offers the solace of what Lee describes as LEGO's attraction: "sheltered, quiet, indoor play" (34). "Indoor" in this context is never really a finalized interior. To apply the example of the LEGO treehouse, the all-consuming architecture leaves little room for us, which, from the time-limited enjoyment of children's access to the LEGO exhibit at Papalote, might even come as a relief.

In sum, the LEGO display at Papalote sets children to solve the transportation problem from without—without the possibility of success—in order to protect the children's sense of safety. It bears emphasizing that if children were able to use a solution of truly sustainable transit, they probably would not have been able to arrive at Papalote in the first place, or at least not at the version of Papalote as we know it.

Árbol Ramón and the View from the Exterior

The endless task of playing so seriously leads me to Árbol Ramón, near the LEGO area on the ground floor of Papalote. The tree is decorated inside and out with replicas of diverse fauna and flora in a lesson on the food chain accessible only by tour. At scheduled times, groups of punctual visitors listen to a *cuate* talk about the animals on Ramón, in a dynamic lecture that my son and I have enjoyed three times; the best tour had the guide hand my son a flashlight so he could illuminate the discussed animal replica. We could never trouble an out-of-doors tree in this way, not only because most real trees lack an opening at the base for tourists to enter but also because our attention, combined with that of our fellow 23 million-plus museumgoers, would undoubtedly doom an outdoor tree to an early demise. In this sense of overpressured resources, I am grateful that Árbol Ramón affords us the opportunity to pretend that we

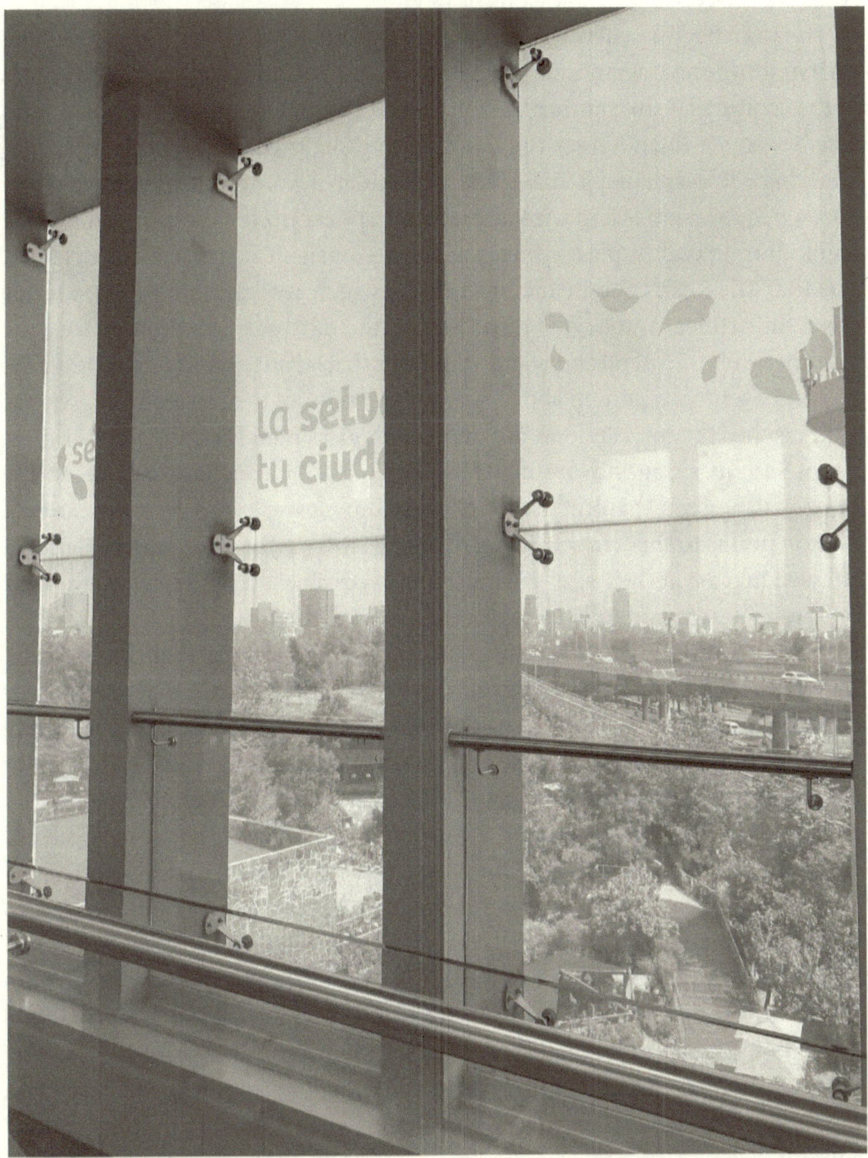

Figure 2.2. View of Mexico City from inside Papalote Children's Museum. Photograph by the author.

are playing with a tree, though I worry that this diorama never successfully counters the constraining lessons of LEGO city.

The tour begins when the *cuate* gives an introduction to the food chain, verifying time and again that almost no one in the audience knows the names of the creatures visible among Árbol Ramón's branches. Then, the host invites listeners to stand and enter the trunk of the tree through an opening by descending a few stair steps. Inside the tree, which accommodates some ten or a dozen people, beetles and similar creatures appear on the interior walls of the trunk illuminated in plastic boxes and with labels of Latinate nomenclature. The tour then exits the inside of the tree, which was still a kind of exterior given the artifice, and begins an ascent on the outside of the indoor tree; the visitors climb the metal stairs and walkways that wind around and around the tree and eventually end up above it, at the top floor of the museum. The visitor who reaches the top can look down through glass panels on the floor and see Árbol Ramón, along with a stuffed bird with outstretched colorful feathers. The nimble visitor can then return to the museum's lower levels on a slide. Árbol Ramón points to the larger trouble, plainly visible on the observation platform encased in glass at the top of the museum, accessible after completing the tour of Ramón and meant to be enjoyed before using the slide.

The worst thing would be not to have a fake indoor tree at all, an idea that I understand from trying to entertain an active child in a city that cannot host children safely in the street. But perhaps the second-worst thing is to rely on the fake indoor tree, because apparently we cannot have the tree without menacing environmental problems. Those problems will be on the educated person's mind as she stands at the museum's top, checks the smog level and car traffic flow beyond the glass, and reads messages stenciled unobtrusively in Spanish on the floor-to-ceiling windowpanes. The vantage of multiple trips to this observation room allows for comparison across seasons, and I must add that the visit over the summer of 2024, during a heat wave, meant the room was so hot that no one stayed and looked out the windows, much less read the stenciling. It is no exaggeration to say that the room was sweltering, already showing the edge of all that glass; we were being cooked in intensified exposure to the actual weather all around us outside by virtue of being indoors and yet suffering worse temperatures.

On less hot days, tourists can linger and read the messages counseling appreciation of plants, which have a "superpower" [súperpoder]: they transform toxic substances into oxygen that "we" [nosotros] can breathe; they trap dust and other harmful substances; they use roots to absorb toxic substances and prevent them from polluting the ground or returning to the air. Additionally, adopting and caring for a plant contributes to clean air in Mexico City. Some stenciled messages end with statements of gratitude: "Thank you, plants!" [¡Gracias, plantas!] and "Thank you for defending us!" [¡Gracias por defendernos!]. If we are so grateful to plants and not just Ramón, there might be a live one in the room, but no, this is LEGO city. The museumgoer stands high above all the treetops in a glass room on the imaginary exterior of the world for discussion, this world of plants, without which we cannot breathe but with which we cannot coexist, at least not in the museum tower, which thus somehow places us again on the exterior of the food chain we are meant to study and reminds me of the vegan-unfriendly food court. The plants didn't disappear. They are just on another level of the exterior.

As if to encourage an engineering solution to this eternal gardening problem of LEGO cities, one stenciled message asks where the viewer would put more trees: "Where would you plant more trees?" [¿En dónde pondrías más árboles?]. Excepting the possibility of installing vertical gardens on the pillars of autocentric infrastructure visible through the windows, the question is rhetorical; from this distance in the tower room, the eye cannot see spaces that lack plants except for the seemingly necessary asphalt and concrete, because how else will people arrive except on the roadways? In a suppression that imitates the lack of concern for roads in LEGO toys, a social media kerfuffle over bike lanes notwithstanding, the signage of Papalote utterly ignores these roads. The spectator is meant to be trained to notice other infrastructural details, as coached by informational panels placed on the floor in front of the windows. To the northeast, one panel explains, the visitor might see the flagpole for the oversized Mexican flag, the Presidente Intercontinental Hotel, and the skyscrapers named Torres Ruben Darío, Torre Mayor, Torre Reforma, Torre Bancomer, and Torre Latinoamericana. Other labeled points of interest on the silhouette represented in the panel with its overly compressed, entirely domesticated horizon include a roller coaster, the National Auditorium, the Pemex Tower, the National Museum of Anthropology, and the monument Estela de Luz. Significantly, as far as I can recall, none of these sights includes a named tree or a named roadway.

Some landmarks more aligned with natural history figure into the panels; the southeast vista of Mexico City purportedly allows a view of the largely undeveloped preserve known as Desierto de los Leones, just as the northeast view supposedly looks out on Chapultepec Lake. However, a good portion of these views may be imaginary, as the viewer lacks the necessary elevation to see all the named elements. As with the LEGO activity, the experience with Árbol Ramón that concludes in the glass observation tower may leave children still on the exterior of an architecture comprised of openings without actual doors, an architecture of insatiable domestication that excludes nature but moves that same nature into an extreme experience, as in the heat event. My reader may object that surely the green areas of Papalote compensate for some of these LEGO-like designs. On half of my visits, the ground-floor pathway outside the museum, open only to Papalote visitors, has been largely out of commission. In the hot May of 2024, the *chinampas* (floating gardens) display, finally open, didn't lend itself to a tour for my children and me because I could not manage to wait in the uncomfortable sunlight. When the plant activities were under construction and since their reopening, from what I could tell, the signs for the garden area encourage visitors to explore and test their senses, and yet owing to the crowds the museum handles each year, children should not leave the pathway. Asking people to use their senses from the pavement flirts with pushing kids ever more consciously into the same old Anthropocene, as they might register the noise from traffic, including the aerial sort, along with smells of smog.

I am not sure that my son and I really climbed a tree if we used a metal staircase, but even so, Árbol Ramón impresses us. Other literature fans might agree, and I note that in the Mexican anthology of tales about Papalote written for children, *A Papalotear* (2013), three authors mention the giant tree by name: Ana García Bergua (33), Fabrizio Mejía Madrid (89), and Julián Herbert (115). Another three writers refer to the tree without using the name: Juan Villoro (42, "an enormous tree, that grew clear to the roof" [un árbol enorme, que crecía hasta el techo]), Martín Solares (126, "the very tall tree that was at the center of El Papalote" [el árbol muy alto que estaba en el centro de Papalote]), and Jorge F. Hernández (146, "a fake tree" [un árbol de mentiritas]). By contrast to these six writers who refer to the tree, only two authors, Vivian Abenshushan (62) and Hernández (146), mention LEGOs, perhaps because LEGOs are available nearly anywhere, unlike Árbol Ramón. The low quality of these tales discourages me from incorporating details into my analysis here, and instead, I move to the last exhibit of interest, on the axolotl, the ajolote, which just happens to correspond to an anthology for adult readers, Roger Bartra's *Axolotiada* (2011). Bartra anticipated by about a decade a mania for the axolotl that inspired sales of innumerable T-shirts and trinkets, such as low-cost stuffed axolotl figures attached to key chains. As recounted in Fornoff's *Subjunctive Aesthetics,* the axolotl surfaced in a tweaked photogenic rendering for Mexico's fifty-peso bill, introduced in 2023 (92).

Axolotl and Museum Games

My son did not pay much attention on any of our visits to the three tanks of live ajolotes; these agile amphibians are nocturnal, and most human observers watch them while they sleep (Cota Hiriart, *Faunologías* 30). My kid prefers a video game on helping the axolotl, largely presented in the singular. This game explains, in vague temporality, that previously ("antes") the ajolote's home looked a certain way, and now its food supply suffers from lilies and trash, while tilapias eat the ajolote's young. The player then has 60 seconds to manipulate the touchscreen and clean up the habitat, using selectively scissors, a net, and a bag. The neatness of this task recalls the clean edges of the LEGO pieces, just as the game title, "Help the Ajolote," echoes the LEGO cue cards in their pleas for solutions. My son took no interest in the low-slung round table in this corner of the museum that presents data on the animal and delivers one sentence per triangular wedge of the table. The information presented on the adult ajolote repeats the entreaty of other activities discussed for Papalote: "Can you help us?" [¿Puedes ayudarnos?]. The review of the ajolote maturation cycle with a round table adroitly presents a wheel of life, passing endlessly from one stage of development to another, which the reader might enter in the middle. This circular design turns the stages of the ajolote into an unending process and not a destination achievable through linear development. The implications ought to please children, or at least adults who can interpret the symbolism and are fans of children, because the relationship of the adult with the younger stages of life seems nonhierarchical.

I find that my son's lack of interest in the exhibit—which gives me little to say about its message for kids—forces me to step outside the bounds of Papalote and draw on literary antecedents that most educated adult visitors will know. The most prominent of these references is Julio Cortázar's short story "Axolotl," from 1956. To enrich my discussion, I will approach that text by way of others in dialogue with Cortázar's piece. Recent contributions by Cota Hiriart and Lemus may not be so familiar. Cota Hiriart's *Fieras familiares* (Family/Familiar Beasts) provides an answer to a question he launches in a book review ("El museo portable"). The opening sentence of his review of *Álbum de plantas prohibidas* (Album of Banned Plants) asks whether it makes sense to conceive of a museum of paper, a question Cota Hiriart implicitly narrows in his book *Fieras familiares* by referring to natural history museums in particular (138, 139). In a conceit sustained by way of section headings, *Fieras familiares* suggests that a book in itself can indeed constitute a museum. I deeply appreciate *Fieras familiares* as a smart, lively read, and my criticism of it does not mean to ignore the pleasure of this interesting text.

The reader understands Cota Hiriart's technique with only a few examples in *Fieras familiares* such as the introduction title, "Gallery Text: Entrance to the Exhibit" [Texto de sala: entrada a la exhibición] (13), and part 1, on a "living museum" of animals in captivity [Museo viviente. Primera parte: Cautiverio] (21). The first species considered is the ajolote, immediately renamed in the

Latinate tradition, in the way of the Papalote displays, as *Ambystoma mexicanum* (25). After listing scientific information about the species, Cota Hiriart follows up with a section labeled "Diorama 1." In captivating autobiographical detail, he reviews such Anthropocenic follies as his early school project of treating animals with hormones to see if the chemicals induced a transformation. In his book review on banned plants, Cota Hiriart employs a neologism for his genre: "literanature" [liternatura] ("El museo portable" 139). This nature is clearly coached from a STEM perspective, and even as no contemporary Mexican writer has published as many pages on the axolotl as Cota Hiriart, he seems to have little to say about his peers' writing on the animal. Where Cota Hiriart excels is in the delivery of factual information on the animal as well as its plight.

In a litany that recalls the video game on the ajolote sponsored by Papalote, Cota Hiriart observes that owing to factors like invasive species, fragmentation, and pollution of the habitat, as well as sales on the illicit market, the numbers of axolotl are dwindling (*Fieras familiares* 48). As a result, Cota Hiriart launches the question Why would we "save"—"salvar," in quotation marks—a species from extinction if the survivors are condemned to live in confinement? (48). This query gets at the heart of the trouble with the LEGO city, as it fosters a kind of confinement that turns out to be conditioned by an exterior that cannot stand in the long run. The problem of extinction that Cota Hiriart treats nevertheless proposes the same paradox as the play instructions in Papalote. If the Anthropocene were to be solved, the tradition of the natural history museum that coaches his *Fieras familiares* would no longer stand either. That is, if Cota Hiriart were to solve the problems he outlines around the ajolote, he would need to rethink the museum itself, along with the scientific tradition, but Cota Hiriart does not do that, with the possible exception of naming a section "Living Museum. Second Part: Freedom (The Islands)" [Museo viviente. Segunda parte: Libertad (Las islas)] (159). There is no exterior to the museum, then, if the animals living freely are nonetheless under institutional protection, such that Cota Hiriart can hire a guide and embark on treks to observe them.

As predicted by this asphyxiating LEGO city design, Cota Hiriart holds few illusions. Even so, he never renounces science. The autobiographical writing in Cota Hiriart's *Fieras familiares* expresses pessimism but not poignant regret; the narrator never seriously repents the experimentation he conducted on ajolotes for early competitive science work. Though Cota Hiriart waxes pessimistic and seems to want to solve the problem of the Anthropocene, he never quite wishes to dispose of the problems of the museum, perhaps because such a solution might endanger his writing under the genre literanature. The reader surely suspects my angle here: Cota Hiriart's other texts on the axolotl include the less captivating *El ajolote,* which features an epilogue that catalogs the ongoing population decline for free ajolotes, across all eighteen types of the species found in Mexico (60). Though it lacks the self-conscious museum play of *Fieras familiares,* the same nonfiction approach of a museum diorama emerges in *El ajolote* as it repeats the astonishing facts about the axolotl related in *Fieras familiares,* including the likelihood that its ability to regenerate perfect body parts has to do not with stem cells, as was previously thought, but with an ability to indifferentiate already existing cells (*El ajolote* 42). As with the acceptance of hormone injections in the name of scientific experimentation, Cota Hiriart never vigorously denounces the ethically dubious practice of cutting off axolotl body parts to watch them regrow, as the autobiographical voice in *Fieras familiares* remembers having done, in excellent detail.

Papalote and its institutional control over visitors almost necessarily fail on the point of freely played imagination, a failure in some ways supported by Cota Hiriart's brief and hardly insightful review of literary texts on the ajolote. His short list of texts on the axolotl, each given scant treatment, suggests that literanature will not have much to contribute to literary criticism, though his bibliography points the reader toward the 2011 anthology *Axolotiada* (Cota Hiriart, *El ajolote* 59 and *Faunologías* 34). The two texts that Cota Hiriart describes in greatest detail are "Axolotl" by Cortázar and "Ambystoma tigrinum" by Elizondo.[2] In the groundbreaking story, Cortázar's writer-narrator visits a zoo in Paris and observes through the glass of an aquarium the axolotl, in a habit so persistent and intense that the narrator becomes that axolotl and the axolotl replaces him as the man who visits the aquarium. Cota Hiriart rehashes the plot but offers no interpretation (*El ajolote* 58). Similarly, Cota Hiriart cites twice—in *El ajolote* and in a book from 2015, *Faunologías*—a quotation from Elizondo's "Ambystoma tigrinum," always without acknowledging the misleading nature of this citation. Cota Hiriart contends that Elizondo kept a tank with several live examples of the species in his home (*El ajolote* 59). This seeming harmony with the animal betrays the actual content of Elizondo's tale, a harmony insinuated by Cota Hiriart's decision to cite, twice, Elizondo's narrator's praise of the amphibian nature of the axolotl (*Faunologías* 34; *El ajolote* 59). Curiously, Cota Hiriart omits the most shocking aspect of Elizondo's text: the narrator's failed attempt to transfer the head of one specimen to the body of another, a gruesome amateur surgery that, given Elizondo's predilections, may not be entirely fictitious. This omission suggests to me a collector's avarice for arranging samples and generating data; to judge from his work thus far, Cota Hiriart is more invested in the maintenance and continuation of the museum than in its destruction, no matter how he might despair for the well-being of the ajolote.

The writer who takes on Cortázar's and Elizondo's legacy more aggressively is Lemus, whom Cota Hiriart also mentions among the authors collected in Bartra's *Axolotiada* anthology. In the story "Larva," published in that 2011 anthology, Lemus reworks what the reader will recognize at first as Elizondo's text: the writer X, increasingly drunk and anxious, decides to play with his pet ajolotes by burning one with a match until it cannot survive and then cutting its body with a scalpel, selected from a surgical instrument collection also kept at home. After the surgery divides the cadaver again and again until the desk is covered in bits of the axolotl, X realizes that any mystery housed in the ajolote's body pertains to a live specimen (Lemus 334–335). Here, the story seems to nod to Cortázar's work as well. X presses his nose against the aquarium and nevertheless fails to see the ajolote as an individual, unique example of the species. The story ends with X cutting a head, apparently his own, a move that the ajolote ignores (Lemus 335). Lemus's tale does not move beyond the natural history museum of Cota Hiriart's interest. The habitat of destruction emerges from the start. Lemus's X possesses a studio with a desk, an aquarium, and a surgical collection, all located in none other than the Coyoacán neighborhood of Mexico City, where the reader may know that Elizondo had a home (Lemus 333). The mention of Coyoacán already invokes the LEGO city, where water is scarce, the air can be bad, the traffic is maddening—and this area of Mexico City denotes real estate of wildly pricey privilege.

Lemus adds another exterior here by trapping the would-be action on the exterior level of intention. Lemus clarifies that the tale of X is only a plan for writing, which necessarily demands contemplation and not vicarious identification from the reader, as it delivers a concluding set of rules. The claim to present a plan rather than the final product conjures a text meant for studied pondering and intimates the appropriateness of the kind of reader reception given a work hung on the otherwise blank wall or papers archived under institutional care. Therefore, in my reading at least, Lemus moves Elizondo's legacy back to the museum, where Cortázar's narrator never left and where Cota Hiriart's narratives prefer to stay. The rules of Lemus's piece stipulate that the planned text will be between eight and twelve pages; literary slang will not replace scientific language; X's nights and activities will not be romanticized; the ajolotes will not be treated anthropomorphically or given symbolic value; and the ending cannot have the narrator enter into the animal's skin and narrate from there (Lemus 336). The unknowability of the ajolote here suggests that possibly not natural history but artful writing—an elite, museum-quality, metaliterary art—best captures what the Anthropocene has so far failed to respect about the nonhuman, non-LEGO world. Unlike the traditional static dioramas of the natural history museum such as Árbol Ramón, the game that Lemus sets up remains in an incipient stage, with a potential to metamorphize into another stage that never occurs, perhaps for the better. Although Lemus's "Larva" coincides with Papalote and the emphasis on rules and goals, the story never suggests a pragmatic goal beyond the always postponed production of a final text.

Lemus's text ultimately hints that safety is impossible from the conditions of the game and that the game is unsolvable under these same rules. The deranged narratives about the axolotl, as Lemus's tale already hints, cannot return the player, the museumgoer, or the literanature fan to the state in which this play would be harmless, a mere contemplation from the exterior of the Anthropocene. To visit Papalote, the Natural History Museum, or one of the many texts on the ajolote is never to remain on a benign exterior in which no animal was harmed in the making, romanticized, or anthropomorphized. Setting children to the task of solving the axolotl game seems, in some ways, especially rigged. Lemus's title, "Larva," refers to the idea that the ajolote never matures into a salamander. Still, immaturity is perhaps part of the relief available to the visitors of these museums and their struggle with presenting a legacy of horrors.

Conclusion

As I wrote a first draft of these words in the summer of 2023, Mexico City and its environs were threatened by a deadly heat wave, volcanic ash from Popocatepetl, the ongoing lash of a pandemic-inducing virus, and other dramatic infelicities already described, like water and transit crises. June 2024, as I said, saw my children and me endure intense heat in Mexico City. Owing to the water crisis, I had considered canceling the trip. We visited anyway and found that amid daily routines, few people mentioned the water problem.

We are hooked on Papalote, in an ongoing relationship that reminds me of the guest at the Hotel California according to the rock band the Eagles' eponymous song, who can check in but never really leave, because the museum is bounded by layers of other exteriors: outside the food court, the parking garage, or the ride-share pick-up, which are in turn bounded by the second section of the Chapultepec Forest, which is in turn bounded by autocentric roads, which are in turn bounded within the megalopolis of Mexico City, itself located in a struggling Mexico, and so on in a chain of doorless and yet entrapping designs. Papalote already compensates for this nightmare, and I will close with this first exhibit.

The oversized creature in the lobby of the museum that visitors encounter after clearing security and fording the glass doors is a papier-mâché "Monstruo come pesadillas." This "Nightmare-eating monster" moves and speaks (plays a recorded message) when children feed a paper into a box. Maybe I am envious of this ritual because I seem to have written my worries and fed them to you. I have largely suppressed the exhaustion for caregivers of the children's museum because such details come across as extraneous. In closing, the nightmare that we ought to process takes into account the constant vigilance prescribed for parents to keep the dangers at bay for the children, even in the museum and of course in all its exteriors, by way of a design that always places kids just outside of an abstract ideal of guaranteed safety. An adult caregiver must accompany children everywhere, meaning that true play, the nonengineered sort, is rarely available. And yet, the fatigue of parenting in the Anthropocene never relents because we never stop demanding that caregivers optimize play such that it resists relaxation and, in a seemingly inevitable side effect, promotes consumption. The more we aim to retreat into that impossible safety, the more we place ourselves on the exterior of it, creating the problem with our solutions, insisting on a world that has no room for us. Here is where we must change.

Notes

1 The most visited section of Chapultepec is the first, which concentrates 66 percent of annual visitors; people tend to arrive there via public transit (Garduño Serrano 352). Circa 2011, Chapultepec received annually "about 15 million visitors," fewer than New York City's Central Park with its nearly 25 million visitors, but "far above any park of its kind in Latin America" [por encima de casi cualquier otro parque de su tipo en América Latina] (Escotto 259). In June 2025, the governmental website for Chapultepec Forest listed the annual visitor total as "more than 24 million" [más de 24 millones] (Bosque de Chapultepec, http://bosquedechapultepec.mx:9000/bosquedechapultepec/). Thus, in about fifteen years, an official estimate adds about ten million more visitors per year.

2 The story title in Elizondo's 1972 publication contains an error with the first "r," as Arturo Ruiz Mautino has recently discussed. Elizondo writes, mistakenly, *Ambystoma trigrinum*, a slip that Andrés Cota Hiriart silently and repeatedly amends in his citation of the story, as does Roger Bartra's anthology *Axolotiada*, thus returning the title to a proper reference to the axolotl that is not in danger of extinction, *Ambystoma tigrinum*. Ruiz Mautino cites Susan Antebi's thoughts from a 2008 article published in *Latin American Literary Review* on Elizondo's seeming lack of differentiation between the species of *Ambystoma mexicanum* and "*trigrinum*" and then engages deeply with this taxonomic history through texts by colonial writers like Bernardino de Sahagún.

Works Cited

Abenshushan, Vivian. "La inolvidable noche de las narices nuevas." In *A Papalotear: 20 años de Papalote Museo del Niño,* illustrated by Manuel Monroy, 55–70. Papalote Museo del Niño, 2013.

Adrià, Miquel. "Legorreta." In *Legorreta Guide,* edited by Miquel Adrià, 6–14. Arquine, 2017.

Anker, Elizabeth S. *On Paradox: The Claims of Theory.* Duke University Press, 2022.

Antebi, Susan. "A Tiger in the Tank: A Literary Genetics of the Mexican Axolotl." *Latin American Literary Review* 36, no. 71 (2008): 75–98.

Bartra, Roger. *Axolotiada: Vida y mito de un anfibio mexicano.* Edited by Gerardo Villadelángel Viñas; anthology, introduction, and notes by Roger Bartra. Instituto Nacional de Antropología e Historia, Fondo de Cultura Económica, 2011.

Cortázar, Julio. "Axolotl." *Final del juego.* Los Presentes, 1956.

Cota Hiriart, Andrés. *El ajolote: Biología del anfibio más sobresaliente del mundo.* Illustrated by Ana J. Bellido. 2016. 2nd edition, Elefanta del Sur, 2022.

Cota Hiriart, Andrés. *Faunologías: Aproximaciones literarias al estudio de los animales inusuales.* Illustrated by Ana J. Bellido. Montzalez, 2015.

Cota Hiriart, Andrés. *Fieras familiares.* Illustrated by Ana J. Bellido. Libros del Asteroide, 2022.

Cota Hiriart, Andrés. "El museo portable de las revelaciones botánicas." Review of *Álbum de plantas prohibidas,* by María del Carmen Tostado Gutiérrez. *Revista de la Universidad de México,* June 2022, 138–141. https://www.revistadelauniversidad.mx/articles/7b5a428a-ead9-4b1e-8437-d1773730ab15/album-de-plantas-prohibidas-de-maria-del-carmen-tostado-gutierrez.

Elizondo, Salvador. "Ambystoma tigrinum." In *Axolotiada: Vida y mito de un anfibio mexicano,* edited by Gerardo Villadelángel Viñas; anthology, introduction, and notes by Roger Bartra, 307–315. Instituto Nacional de Antropología e Historia, Fondo de Cultura Económica, 2011.

Escotto, Daniel. "Un plan para la segunda sección del Bosque de Chapultepec. A Plan for the Second Section of Chapultepec Forest." In *Bosque de Chapultepec,* translated to English by Debra Nagao, 259–270. Horz, 2011.

Espinosa López, Enrique. *Ciudad de México: Compendio cronológico de su desarrollo urbano (1521–2000).* Instituto Politécnico Nacional, 2003.

Fornoff, Carolyn. "Reflexive Extractivist Aesthetics." *FORMA Journal* 2, no. 1 (2023): 37–69.

Fornoff, Carolyn. *Subjunctive Aesthetics: Mexican Cultural Production in the Era of Climate Change.* Vanderbilt University Press, 2024.

García Bergua, Ana. "La rana equilibrista." In *A Papalotear: 20 años de Papalote Museo del Niño,* illustrated by Manuel Monroy, 25–37. Papalote Museo del Niño, 2013.

Garduño Serrano, Blanca Mónica. "El Bosque de Chapultepec: Espacio público de la capital en tiempos de urbanismo neoliberal." In *Bosque de Chapultepec,* translated to English by Debra Nagao, 337–389. Horz, 2011.

Gordon, Doug, and Marcel Steeman. "Where Are the Bike Lanes in Lego City?" *The War on Cars* podcast, episode 65, June 7, 2021. https://thewaroncars.org/episode-65-where-are-the-bike-lanes-in-lego-city-final-web-transcript/.

Grosz, Elizabeth. *Architecture from the Outside: Essays on Virtual and Real Space.* MIT Press, 2001.

Harari, Yuval Noah. *Sapiens.* HarperCollins, 2015.

Herbert, Julián. "Bernardo, el hijo del ratón." In *A Papalotear: 20 años de Papalote Museo del Niño,* illustrated by Manuel Monroy, 107–122. Papalote Museo del Niño, 2013.

Hernández, Jorge F. "Papalote de noche." In *A Papalotear: 20 años de Papalote Museo del Niño,* illustrated by Manuel Monroy, 139–146. Papalote Museo del Niño, 2013.

Jørgensen, Dolly. "Portraits of Extinction: Encountering Extinction Narratives in Natural History Museums." In *Traces of the Animal Past: Methodological Challenges in Animal History,* edited by Jennifer Bonnell and Sean Kheraj, 371–387. University of Calgary Press, 2022.

Jørgensen, Dolly. *Recovering Lost Species in the Modern Age: Histories of Longing and Belonging.* MIT Press, 2019.

Lee, Jonathan Rey. *Deconstructing LEGO: The Medium and Messages of LEGO Play.* Palgrave Macmillan, 2020.

Lemus, Rafael. "Larva." In *Axolotiada: Vida y mito de un anfibio mexicano,* edited by Gerardo Villadelángel Viñas; anthology, introduction, and notes by Roger Bartra, 333–337. Instituto Nacional de Antropología e Historia, Fondo de Cultura Económica, 2011.

Mejía Madrid, Fabrizio. "Calcetín." In *A Papalotear: 20 años de Papalote Museo del Niño,* illustrated by Manuel Monroy, 89–94. Papalote Museo del Niño, 2013.

Papalote Museo del Niño. *Informe anual 2014.* https://www.papalote.org.mx/wp-content/uploads/2023/10/Reporte-2014.pdf.

Papalote Museo del Niño. *Informe anual 2023.* https://www.papalote.org.mx/wp-content/uploads/2024/06/Reporte-2023.pdf.

Ruiz Mautino, Arturo. "Experiments on Fantastic Biology: Salvador Elizondo's Axolotls and the Colonial Trace." Modern Language Association Annual Conference, 10 January 2025, New Orleans.

Servitje de Lerdo de Tejada, Marinela. "Papalote Museo del Niño. Papalote Children's Museum." In *Bosque de Chapultepec,* translated to English by Debra Nagao, 121–125. Horz, 2011.

Solares, Martín. "Norah y el tiburón." In *A Papalotear: 20 años de Papalote Museo del Niño,* illustrated by Manuel Monroy, 125–136. Papalote Museo del Niño, 2013.

Vázquez Martín, Eduardo. "El museo de historia natural y el tiempo que vivimos. The Museum of Natural History and Our Times." In *Bosque de Chapultepec,* translated to English by Debra Nagao, 111–114. Horz, 2011.

Villoro, Juan. "La nube embotellada." In *A Papalotear: 20 años de Papalote Museo del Niño,* illustrated by Manuel Monroy, 39–52. Papalote Museo del Niño, 2013.

3

A Contemporary Cabinet of Curiosities

Play, Improbability, and the Reinvention of the World in Liliana Porter's *El hombre con el hacha y otras situaciones breves*

JERÓNIMO DUARTE-RIASCOS

A Splendor of Destruction

It is difficult to escape chaotic enumeration when attempting to describe Liliana Porter's installation *El hombre con el hacha y otras situaciones breves* (Man with Axe). The piece, which became widely known after its inclusion in the Pavilion of Time and Infinity during the 2017 Venice Biennale *Viva Arte Viva*, features an improbable cast: a man with an axe; shattered figurines of cultural icons (Mao, Mickey and Minnie Mouse, JFK, El Che, Napoleon); shattered figurines of generic characters (humans, animals, animated beings); clocks, broken clocks, toys, chairs, mirrors, Coca-Cola bottles, cannons, boats, postcards, books, hammer and sickle, cars, a piano (figure 3.1).

The installation's centerpiece is a minuscule man who, despite his size, commands the entire space.[1] The spectator is faced with what Porter has identified as a "splendor of destruction" that emanates from the figurine's axe. The closer the object is to the man and his axe, the more destroyed it is. The further the object, the larger its scale. It is unclear if the man is beginning or concluding his destructive spree; pulverized debris whose origin is difficult to assess can be found right under his axe, while a full-size disemboweled wooden piano is located at the other end of the room.

In this essay, I will argue that *El hombre con el hacha* exposes the improbable certainties that sustain our relation to others and to the world. To do so, the installation functions in a way similar to that of historical cabinets of curiosity, deterritorializing the spectator and taking them to another time and place. Yet, instead of materializing the certainty of an exotic or removed alterity, Porter's contemporary cabinet of curiosities collapses certainties and promises a revelation that is always already deferred, in a nod to Borges's understanding of the aesthetic act as, precisely, the imminence of a revelation.

The piece alludes to historical cabinets of curiosity, more commonly used in natural history, to reinterpret this exhibition apparatus from a contemporary vantage point. The perspective shift operates a critique that exceeds these devices of knowledge and extends to (mostly Western) claims of certainty about the world and the beings that populate it. Porter's *El hombre con el hacha* creates awareness about the artificiality of meaning that was absent from the experience of historical cabinets of curiosity, and thus the work becomes a tool to better understand the effects of similar mechanisms of knowledge production in the construction, use, and abuse of ideas of alterity.

To support these claims, I first illustrate how *El hombre con el hacha* synthesizes many of Porter's preoccupations throughout her career and functions as a unique retrospective of her practice. Then, I discuss the peculiar temporality the piece creates and its role in the displacement it elicits. Finally, through a more detailed analysis of the parallels between *El hombre con el hacha* and historical cabinets of curiosities, I demonstrate how Porter exposes the improbability of certainties while at the same time giving the spectators tools to materialize a reinvention of the world.

Before Venice, the installation had been exhibited twice, in 2011 at the Museum of Arts and Design in New York City and, three years later at the Museo de Arte Latinoamericano de Buenos Aires. In all three instances, Porter has stated that the installation can be read as a retrospective of her practice, featuring recurring motifs and themes that have preoccupied the artist for decades. The claim is easily sustained. There are overused images that question and reveal processes of meaning construction; plenty of small vignettes that stage improbable, often contradictory, situations; and a simultaneous cohabitation of "things" that belong to different ontological, epistemological, and phenomenological registers.[2] There is also a meditation on memory and time, and the piece thematizes what is perhaps Porter's central preoccupation: the tensions between reality and representation and the possibility of the latter superseding the former. Commenting on this matter in a 1991 interview with Mari Carmen Ramírez, Porter states, "Ultimately, my claim is that there is no such thing as objective reality or things as they are. All that exists is our relation to things, and because of this, what we do is always autobiographical in one way or another" [Yo en el fondo sostengo que la realidad objetiva o las cosas como tal no existen. Solamente existe nuestra relación con las cosas, y por eso lo que

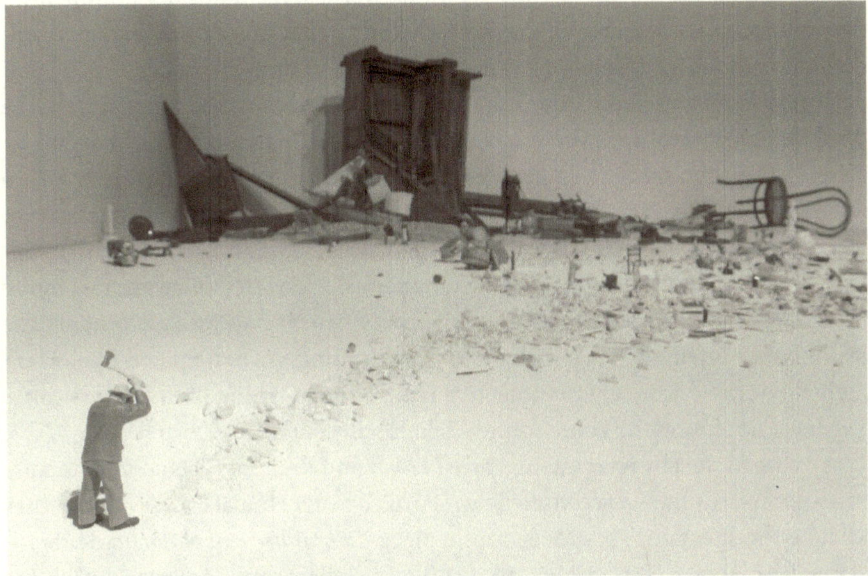

Figure 3.1. Liliana Porter, *El hombre con el hacha,* 2017. The installation was on view at the Venice Biennale *Viva Arte Viva*, among other venues. Photo credit: Liliana Porter.

hacemos siempre, de algún modo va a ser autobiográfico] (in Ramírez, "Entre espejos").[3]

This relational approach is clearly visible in some of her other projects that are also featured in *El hombre con el hacha.* For instance, *Forced Labor,* a series the artist started producing in the early 2000s, is composed of installations where a minuscule figurine—not unlike Venice's little man—embarks on a titanic task. Porter introduces the spectator to weavers, painters, sweepers, and construction workers who are painstakingly devoted to an impossible undertaking. As many of the artist's works, these are pieces imbued with a tragic-comic affect; the characters' faith in their ability to complete the task is in stark contrast to the knowledge the spectator possesses that signals to the futility of the effort. Some of these workers reemerge in *El hombre con el hacha* as if trying to do their part and help to mend the world that their peer is reducing to ruins with his axe.

The thematizing of the construction/reconstruction tension is also at the core of another of Porter's series, aptly titled *Reconstrucciones.* In these works, a small toy figurine stands proudly whole next to a photograph of its shattered version. The dialogue between the object and its destroyed representation is puzzling because it presents, among other issues, a nebulous temporality. Are we, as spectators, witnessing the pristine reconstruction of an object once broken, or is the piece foretelling an inevitable loss of unity, of being?

There is an implied "order" that informs the *Reconstrucciones* series. The broken objects featured there have traces of a unity that one could aspire to. There is no such thing in *El hombre con el hacha.* The splendor of destruction displayed in this installation invites (perhaps even forces) the spectator to propose, to invent, their own way of organizing chaos and structuring meaning. This demand from the piece simultaneously activates an awareness in the spectator, the knowledge that the process of sense-making that renders the installation intelligible is largely their own crafting. As anticipated by Porter's belief that there is no extant objective reality, the meaning of *El hombre con el hacha* is not found; it is constructed relationally.

I chose to start by referencing *Forced Labor* and *Reconstrucciones* not because they necessarily have a preeminence in Porter's oeuvre but because together they dictate the spectator's response to the piece. *El hombre con el hacha* demands a conceptual reconstruction, and just like the characters of *Forced Labor,* the installation leads (tricks?) us into believing that we will get the job done. We are led to believe that once we have labored enough, we will access the revelation the piece promises. Before that, however, the spectator needs to confront the bewildering temporality the installation enacts.

A Frozen Time

The little man's destructive binge in *El hombre con el hacha* surpasses and manipulates time in dual fashion. First, his axe respects no chronology or diachroneity—both the historical and the contemporary simultaneously collapse to his enterprise. Second, this unique and incongruous temporal grasp is presented frozen. Since early in her career, Porter has played with time in both of these ways, creating situations in which different times coexist and experimenting with ways to stop time and simultaneously convey movement.[4]

Here, unlike in other works, the temporal freeze sheds light on the utter chaos of the scene: absence of linearity, simultaneous absurdity, and conflicting affect. Typically this chaos has been read as "evoking something close to the visual workings of memory" as a metaphor for the passage of time and the mnemonic techniques displayed by the human brain to capture the past (Ramírez, "Illusive Fragments," 28). The installation features objects that, on the one hand, refer to a lost childhood and, on the other, constitute icons of public life such as pop culture, art history, and politics. In *El hombre con el hacha* the private and the public coalesce into an unstable site of signification that is merely suggested and seems to be lying dormant under the debris.[5]

I want to offer a slightly different interpretation of the chaos, one that puts Porter's elucubrations in dialogue with some ideas articulated by Argentine artist Luis Felipe Noé in the 1960s, the decade during which Porter moved to New York City and a formative time for her artistic inquiries.[6] Noé, slightly older than Porter and a respected figure in the Latin American artistic community of the time, discusses many of his theoretical concerns about artmaking in a text he titled *Antiestética,* published in 1965. There, Noé proposes "to create an aesthetic of the anti-aesthetic [that would] go against all little orders and try to express an organic vision of chaos and disorder, since these are the main themes of our society" [elaborar la estética de lo antiestético es ir contra todos los pequeños órdenes y tartar de dar una vision orgánica del caos y el desorden, pues éstas son las notas principals de la sociedad en que vivimos] (57).

Noé was writing during the convulsing mid-1960s, and in his assessment of his context, he was mainly concerned with the loss of unity or perhaps more accurately, with unity's destruction—with the ways in which the political, technological, and cultural events of the time were effectively crushing the individual's ideas of wholeness. For him, it no longer made sense to speak of a single focal point because the experience of the real was shattered, responding to a plethora of stimuli coming from divergent directions that chaotically invaded his environment.[7]

Porter, I think, perceives this chaos and puts it on display in *El hombre con el hacha.* However, her chaos does not respond to a specific historical moment but is rather indicative of humanity's relation to the world. As Porter has remarked, "The entirety of my oeuvre is founded on the feeling that one understands nothing, that there is no clarity. Therefore, all my works are rather questions. . . . If you formulate a more precise question, then you move forward, but that doesn't mean that there is an answer" [Toda mi obra parte de la base de sentir que uno no entiende nada, que no la tiene clara. Entonces, todos son más bien como interrogantes. . . . Si uno hace la pregunta un poco más precisa, avanza, pero no porque haya una respuesta] (in Grinblatt). It is from the humility of this vantage point that her work is erected.

Intelligibility and order are notoriously intertwined. Both of these characteristics are absent in *El hombre con el hacha,* yet this is a piece that allows the spectator to move forward in an invented path toward understanding. The piece presents chaos using a multiplicity of resources that have been constant in Porter's practice. First and foremost of those is the fragment. Rather than appearing as part and parcel of a given whole, the artist's fragments are more like ruins, fictions (representations) of the real exuding fragility and multiple temporalities. Art historian Charles Merewether has observed about Porter's work, "In the field of allegorical intuition the image is a fragment, a rune. What remains quietly compelling in Liliana Porter's work is the sense of frailty, ephemerality and transience in the constitution of her images and subjects" (41).

This transience that Merewether identifies and is arguably antithetical to art's pretenses of permanence and immortality is at the very core of Porter's practice. Precisely by letting the objects speak and directly interpellate the viewer, Porter bestows the spectator with the task of engaging with these ruins of the world to create an order that would make them intelligible. That act of receptive creation is materialized through the activation of a childlike imagination. The toylike quality of many of her objects added to a belief in a shared experience that influences, without determining, their meaning, activates the spectator's imagination, and takes them to a different time and space. In that time and space of fiction, many of the limitations of the real simply do not apply; it is a place where, for instance, time can be frozen.

Porter has claimed that she wishes that her work would create an effect akin to that produced when one is engrossed in watching a movie and, suddenly, someone turns on the lights.[8] The experience of the fiction on display is briefly interrupted, altered, yet not fully forestalled. If at any given point the movie is captivating enough to make its viewers forget that they are witnessing a representation, the lights coming on correct that lapse of memory. The lights do not necessarily curtail the spectator's enjoyment of the film or even their engagement with it. They do, however, introduce an awareness that highlights the constitutive fictionality of the experience. Like these lights, Porter's works take their spectator out of the illusion of certainty about the world and underscore the artificiality of meaning.

Improbable Certainties

In the 2017 Venice Biennale catalog, Porter describes her practice: "In spite of the variety of media, there is a common denominator to most of my work, that is a reflection on the interpretation and the construction of reality, and its improbable certitude. . . . I frequently use a cast of found inanimate objects, figurines and toys, to evidence our role as creators of meaning" (in Macel 477). This eloquent proposition, according to which the real is experienced via improbable certainties, is articulated visually in *El hombre con el hacha* and constitutes one of the reasons that make the installation compelling as a piece to reflect on the conditions of contemporaneity.

Throughout her career, Porter has insisted that there is no linearity or absolute order informing one's rapport with reality. Instead, for her, part of the human experience depends on an ability to invent, in an often arbitrary fashion, multiple orders always informed by biography: "I think that the key term here is *to invent* because, in reality, there is no absolute order, there is just the order one imposes on things. And this created order is, certainly, a fantasy" [Creo que realmente la palabra clave aquí es *inventar*, porque en realidad no existe ningún orden absoluto, sino el que uno le adjudica a las cosas. Y el orden que luego se crea es en realidad una fantasía] (in Ramírez, "Entre espejos" 102).

This invention of order is central to the effects and interpretation of *El hombre con el hacha*. The piece is unequivocally simultaneous. As such, it can only be accessed fully as an experience, limited in time and place. Most of Porter's spectators were never able to see the installation live and could only attempt to grasp it through narrative. This later access is necessarily artificial because it responds to an organization absent in the piece. The work's simultaneity questions language's reliance on temporality and exposes the inherent limitations of sense-making when applied to stimuli that resist ordering. Paradoxically, perhaps, this resistance to order makes ordering irresistible. In Porter, said organization is materialized through an act of imagination. Using familiar objects activates the inventive process and takes the spectator to a different time and place. There, imagination functions in the way theorized by Thomas Hobbes as a "decaying sense," in which the encounter with the familiar object triggers an attempt to displace the spectator and take them to the reality where that object was first experienced. In Porter, this attempt is always fraught, yet imagination's failure to restore the real propels us elsewhere.

I will return to the implications of this displacement in a moment, but before that, I want to stress a point I find crucial when approaching the artist's practice. In her works, Porter does not lament humanity's reliance on imagination or denounce the artificiality of meaning, hoping to restore ideas about authenticity. Instead, she thematizes these phenomena from a perspective that tends to negate the postmodern lens. Ramírez notes this clearly her text "Illusive Fragments: Liliana Porter's Art of Memory":

> The logic of Porter's art also embodies a childlike attitude that is ultimately at odds with the nihilistic intent of post-modern production. The playing up of the simulated and the real that characterizes her work has more to do with the ludic quality of the child's experience than with

> the rational critique of representation. Likewise, the impulse to collect fragments and bits of the material world evokes the experience of the child, for whom every bit of discarded object or material can function as an emblem of a fantasy world. (16)

For Ramírez, this childlike quality parodies the seriousness of the adult world and grants a subversive character to Porter's practice. While I substantially agree with this reading, I think there is room for an important clarification. In my view, Porter's fantasy world is subversive not because it opposes the prosaic world of adults but because it illuminates and exposes their utter similarity. Her work humorously emphasizes this kinship (turns the lights on, so to speak), noting that the certitude of the ostensibly real adult world is, at its best, improbable.

I have been arguing that the awareness of this improbability does not prevent access to alternative chronotopes. On the contrary, Porter's works are designed to relish that awareness, activating a childlike imagination that freely transports the spectator to worlds and times that are not those of the literal encounter with the artwork. *El hombre con el hacha* speaks loudly to this effect. I find that this installation in particular functions as a contemporary cabinet of curiosities, presenting fragments that point to and promise access to another reality without ever really fulfilling that promise.

A Contemporary Cabinet of Curiosities

Frequently studied as the precursors of modern museums, cabinets of curiosities (*Wunderkammern*) were popular sites of display during the late sixteenth and the seventeenth centuries in Europe.[9] The objects featured in them, typically divided into four categories—*artificialia, naturalia, exotica,* and *scientifica*—were meant to incite a variety of feelings and give a glimpse into a time and place where a radical alterity resided.[10] Triggered by fragments of an implausible reality, spectators of the cabinet needed to perform an act of imagination that would construct an otherness that often secured the perceived and believed wholeness of the European self.

The categories organizing the historical cabinets largely responded to a desire for "keeping and sorting the products of Man and Nature and . . . promoting understanding of their significance" (Impey and MacGregor xvii). This keeping and sorting impulse, often elicited by an unintelligible reality, provided a structure of order from which to understand not only the objects on display but also the worlds from which they originated. Needless to say, the selection criteria used to populate the cabinets did not provide anything close to a comprehensive sample of these worlds. Instead, and responding less to a scientific urge, the objects in the *Wunderkammern* were imbued with affect and personal connections to the collectors' fancies.

Like those in historical cabinets, the objects on display in *El hombre con el hacha* are emotionally charged both from the artist's and the spectator's perspectives. Unlike the original *Wunderkammern*, however, Porter presents the objects in disarray, with no significant guidance as to how to enact the imaginary deterritorialization they elicit. One could arguably organize the fragments and the little vignettes they compose to fit the traditional categories listed above. However, the hoary toylike quality that characterizes Porter's cast uncannily resists this nominally objective approach. Instead, the ludic and familiar affect that is present in her objects guides the imaginary displacement that every encounter with a cabinet of curiosities aspires to generate. Historical cabinets and Porter's installation share an interest in fragments, ruins even, of foreign times and places. These fragments point to and promise access to a different reality that is never fully apprehended—in Porter, full apprehension is quickly revealed as a fallacy; historical cabinets, instead, tend to disguise this impossibility with the introduction of perceived certainties.

In other words, Porter's contemporary cabinet of curiosities plays with the human drive to keep and sort the world and feeds our desire for understanding and intelligibility but frustrates it from the get-go. The little man with the axe and the many "other brief situations" [otras situaciones breves] on chaotic display incite the spectator's curiosity and trick them into attempting to figure out what is happening. For some, the piece quickly obstructs their desire to organize and comprehend. Once the spectator experiences the installation's impossible apprehensibility, the only option is to navigate it adrift. This is precisely Graciela Speranza's point when asking, "How can you describe the experience without constraining the freedom of the gaze that can wander through the pieces without measure, without destination, without narrative?" [¿Cómo describir (la experiencia) sin violentar la libertad de la mirada que puede vagar por las piezas sin medida, sin rumbo, sin relato?] (12). I tend to think, however, that even when experiencing the installation adrift, the affect that informs and animates the objects pulls the spectator toward a particular order. Certainly, this order (like the affect) is spectator-dependent and consequently deeply subjective and relational; these are characteristics that, according to Porter, also guide our approach to the world.

Cabinets of curiosity, both historical and contemporary, seek to establish and present the position of humans in the larger scheme of things. In dialogue with this quest, historical cabinets were populated by displays of technical virtuosity, frequently materialized in miniature objects that demonstrated impressive feats of the human species. Porter's contemporary cabinet presents feats of its own.

The plethora of brief situations exhibited attests to the most absurd and tragic-comic human deeds: the impossible tasks of forced labor; the relentlessness of a minuscule gardener devoted to watering the image of a flower; a beheaded Mao; and an inexplicably small yet enormously destructive man with an axe, to name only a few.

Historical *Wunderkammern* granted their spectators a glimpse of a time and a place where a radical alterity resided. In Porter, there is also a foretaste of an alternative chronotope in which normalized ideas about time and space are upended. The improbable situations that compose *El hombre con el hacha* disregard spatiotemporal propriety and give access not to a radical alterity but to a rarified self, a self that is made aware of the contingency of its artificial construction through the contemplation of objects that feel familiar even if they are not fully so.

As Ramírez puts it, when confronted with Porter's objects, "we cannot help but be drawn in as participants of an illusive spectacle where childhood heroes stand side by side with utopian referents, where memory displaces experience and proclaims itself as the real" ("Illusive Fragments" 14). Perhaps more than memory, the force that displaces the real, in Porter and the traditional *Wunderkammern,* is imagination. Certainly, the difference is a small one, yet inventiveness remains the only way to materialize a memory of a reality that was actually never fully experienced.

Ramírez's claim was also a preoccupation in the conceptualization of *Viva Arte Viva,* the 2017 Venice Biennale where *El hombre con el hacha* was shown. Christine Macel, curator and artistic director of this iteration of the event, claimed in the exhibition's catalog, "Art is the ultimate ground for reflection, individual expression, freedom, and for fundamental questions. . . . After all, art may not have changed the world, but it remains the field where it can be reinvented" (16–17). Porter's work thematizes this reinvention. When seriously engaging with *El hombre con el hacha* as an independent piece that can also function as a retrospective, it is worth asking what kind of reinvented world the artist is proposing through her practice.

I fundamentally see three different ways to answer this question. I have alluded to the first by acknowledging important scholarship on Porter's work that precedes mine. From this perspective, the reinvented world is that of memory, a space and a time when remembrance replaces experience and a new reality emerges, proclaiming itself as such. Importantly, in this reinvented world, the certainty of linear time collapses, giving way to affective simultaneity.

The second response is perhaps slightly more literal. *El hombre con el hacha* is arguably a piece that functions as a multidimensional critique of the West. Porter herself has voiced her disinterest in this interpretation; nevertheless, I find it worth mentioning because it is very much in dialogue with recent preoccupations not only in artmaking but also in theory and criticism in the humanities. To begin, the installation humorously reminisces on the highly problematic Western practice of human exhibitions, where inhabitants of perceived exotic lands were displayed to the marvel of the European public.[11] The individuals on view were often accompanied by flora, fauna, and native artifacts that would eventually make their way into cabinets of curiosities and later, museums.

Porter's installation plays with this concept, even if unintentionally. After all, what is on display is a little man, notoriously European, in typical Western attire, equipped with an axe and generating a gargantuan destruction. Various mementos and ruins of his decaying world accompany him, indicating that they may have also succumbed to his enterprise. In this case, the world needs reinvention quite literally because this little man reduced it to shambles. The rarity of a tiny individual causing such massive destruction of his own world is undoubtedly remarkable and, for a large portion of the non-Western world, truly exotic.

This reading of the piece can easily draw connections to many contemporary efforts in theory and artmaking that actively distance from and critique the Western tradition and the devastating effects of many of its ventures. I cannot do justice here to the breadth and complexities of these critiques, nor is it critical to the point I am advancing. However, I want to note that a common desire in this trove of reflection and inquiry is that of looking elsewhere—in time, space, premises, and goals.

This desired gaze, in turn, propels the action in different forms, and consequently, the degree of reinvention of the world they propose ranges from blanket change to more moderate departures from the capitalist model.[12] In any case, all these efforts share a resistance to the modernist belief that the Western project constitutes the only correct and desirable way of living. In that vein, authors and artists who endorse this critique see the world not as a unitary, coherent, and cohesive whole but rather as a pluriverse, containing multiple worlds with varied cosmologies, epistemologies, and ontologies.

All of this is very much in dialogue with Porter's work. The improbable encounters that are typical of her practice—say, Minnie Mouse enamored with Ernesto Che Guevara in *Minnie/Che,* 2003 or a penguin in deep conversation with Jesus in *Dialogue (with Penguin),* 1999—suggest and imagine different logics, relationships, and ways of being. More specifically, the nonsensical destruction led by the man with an axe can very well be understood as a meditation on the contemporary state of the Western project with its self-induced ecological and humanitarian crises. As spectators of the exotic civilization on display in *El hombre con el hacha,* one can position oneself directly inside the destruction: pulverized or beheaded figurine or survivor of the wreck, trying to mend the pieces and (re)invent the real, watering the flowers, sweeping the rubble, and so forth. Alternatively, one could position oneself from afar as a curious observer intrigued by the unintelligibility of the debris who uses the ruins as propellers to a different time and place. The rarified self that emerges from the installation to lead said displacement is not prescriptive. Unlike some of the contemporary cultural artifacts that address current crises caused by the imposition of modernity's (improbable) certainties, *El hombre con el hacha* is not a work that points to specific actions in order to condemn or celebrate them. Rather, it lets the spectator imagine what other types of relations it may promise, elsewhere. Porter provides tools to answer questions but shares no specific responses to the issues her practice raises.

Consequently, I think, the artist is more interested in this later elopement than in the critique I just alluded to. She has argued that what truly intrigues her are the metaphysical implications of the confrontation with chaos and not so much its ecological or political dimensions. I identify this metaphysical rumination as the third way of understanding the reinvented world that can emerge from *El hombre con el hacha.* It is true that sensu stricto, the piece does not propose any specific reinvention. Nevertheless, it provides the grounds for such reinvention to occur. Unlike the critical trends I referenced that intensely seek other ways of knowing, I see Porter's work as an invitation to dwell in the discomfort of not knowing.

Even when the end result of an engagement with Porter's practice is the blunt negation of certainties, this is not her starting point. Negating a certitude requires, paradoxically, certainty, and Porter is less concerned with negating certainties and more committed to stressing their improbability. She turns the room's lights on but does not turn off the movie, if you will. Neither does she abrasively mock understanding. The strength of Porter's work lies, as I see it, in how she playfully courts the spectator and invites them to actively engage and produce (invent, reinvent) meaning. Of course, the reinvention of the world Macel talks about is a titanic task—potentially impossible, frequently fraught. Porter is keenly aware of this, and her spectator eventually comes to that realization too.

However, this awareness does not preclude the free attempts to invent our metaphysical (and physical) dwellings and the meaning that can be extracted from them. To reiterate, I do not claim that Porter's practice in general or *El hombre con el hacha* in particular are, in fact, reinventing the world, but they do give access to that possibility. By doing so, claims and attempts to fully understand the unknown and access it objectively or scientifically (such as those proper to the historical cabinets of curiosities) are presented as ludicrous.

Porter does not present meaning; she merely shows one of the multiple ways in which meaning arises: unusual connection, spontaneous imagination, childlike freedom and play, atemporal and nonspatial relationality. Her spectators do not access a new reality fashioned by the artist, yet they do receive tools and techniques to fashion it themselves. Consequently, the construction process becomes conscious, and any claim of certainty resulting from the use of those techniques or tools can only be exposed as flawed and, at times, ridiculed. In Porter's universe, to claim certainty of the real is as ludicrous as correcting a scribble.

To Correct a Scribble

I want to end with a comment about Porter's 2007 piece *The Correction (Red)* (figure 3.2). The drawing is part of a line of her practice that started in the early 1990s and explores, precisely, the implications (and sometimes the absurdity) of correcting. It features a scribble, *un garabato,* with a correction in red superimposed. Talking about the piece and its significance, Porter explains,

> The idea is that it is impossible to correct a scribble, because a scribble, one assumes, is the only thing that, conceptually speaking, cannot be wrong. What interests me about this is that correcting is very similar to what we do all the time when we are trying to grasp something, searching for a theory for things, which is actually sort of funny because it is simply impossible. What I mean to say is that we never know what the final result will be, and our obsession with knowing ends up being a kind of involuntary exercise in humor. Another thing that fascinates me about the scribble is that it reminds me of school, when they would correct something in math that I didn't understand (and I seriously didn't understand), with that steady, sure hand that was so . . .

> La idea se centra en la imposibilidad de corregir un garabato, porque se supone que es lo único que conceptualmente nunca puede estar mal. De esto lo que me interesa es que corregirlo se parece mucho a lo que hacemos todo el tiempo al tratar de aprehender algo, buscando una teoría para las cosas lo cual, por ilusorio, es siempre un poco humorístico. Lo que quiero decir es que nunca sabemos cuál es el resultado final, y nuestra obsesión por conocer termina siendo como un ejercicio de humorismo involuntario. Otra cosa que me cautiva del garabato es que me hace recordar la escuela, cuando te corregían esa tarea de matemática que uno no entendía (en serio no entendía), con esa seguridad del trazo tan . . . (in Porter, Katzenstein, and Volk 169)

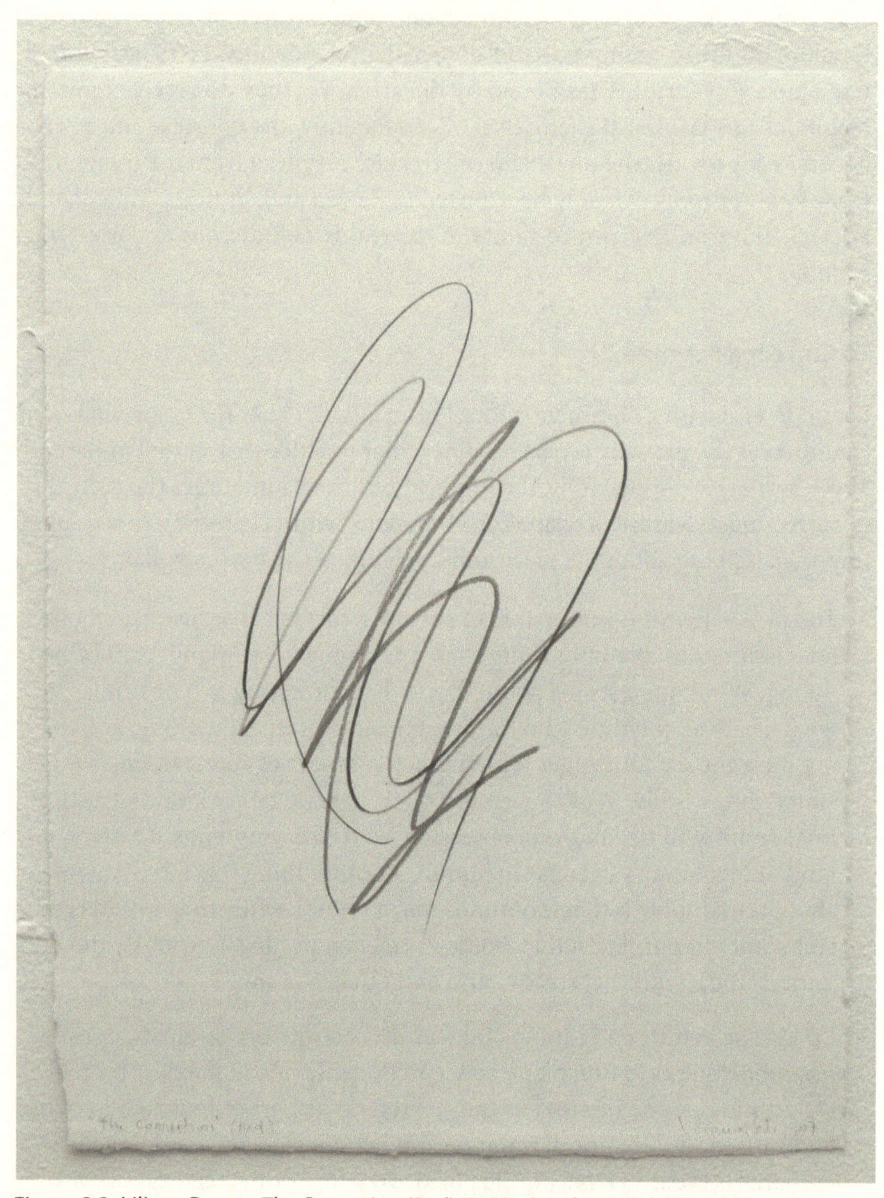

Figure 3.2. Liliana Porter, *The Correction (Red)*, 2007. Graphite and color pencil on paper. Photo credit: Liliana Porter.

I see a parallel between *The Correction (Red)* and *El hombre con el hacha* because the installation triggers this desire to apprehend meaning and organize chaos in a way that is, to an extent, risible. Porter's work exposes, time and again, that any order that allows for intelligibility is a fantasy, a fiction, frequently constructed through acts of childlike imagination.

I have tried to show that none of this is infused with a pejorative connotation. On the contrary, there is something very freeing in the awareness that the real is nothing more (and nothing less) than a chaotic display of stimuli that we, humans, seriously struggle to understand. The experience is not unlike that of the child who grapples with an incomprehensible math problem. Porter's practice does not solve the quandary, let alone attempt to correct it. However, the piece goes one step further, moving away from unintelligibility.

I think there is something to say about one of the most prominent objects in *El hombre con el hacha,* perhaps the hardest to destroy, being the piano, an object I interpret as a metonym for the ineffability of music. Its survival is reminiscent of one of Borges's statements (and Borges has long been Porter's privileged interlocutor): "Music, states of happiness, mythology, faces molded by time, certain twilights and certain places—all these are trying to tell us something, or have told us something we should not have missed, or are about to tell us something; that imminence of a revelation that is not yet produced is, perhaps, the aesthetic reality" (5). *El hombre con el hacha* does not reveal anything, it does not present a reinvention of the world, a correction of it. It does tend to suggest that *a* revelation is imminent. Ambiguous and unclear, this revelation requires patience and the willingness to dwell, for now, in the not-knowing. While waiting, in the frozen time of Porter's work, we are returned to that childhood memory, when we stared at the corrected scribble, hoping that eventually the math homework would make sense.

Notes

1 Porter describes the man as follows: "I put him on the floor in my studio and I saw how much space he took up, even though he is so small, he occupied a great deal of space" (in Grinblatt).

2 This cohabitation is akin to that characterizing cabinets of curiosities. Porter, however, refuses to categorize the objects she displays, thus complicating intelligibility—taking distance from a practice that was central to devices used in natural history such as cabinets.

3 Unless otherwise indicated, all English translations are my own.

4 Since as early as 1968, Porter has been experimenting with ways to visually communicate frozen time. A notable example of this preoccupation is the 1968 series

Wrinkle. Composed of ten photo-etchings that show a wrinkling sheet of paper, the works thematize photography's ability to freeze time. Additionally and because it is a series of ten images, when shown together they succeed in conveying movement, yet they do so precisely by freezing it.

5 Discussing this coalescence between the public and the private, Ramírez continues, "Porter's icons, however, are more than personalized mementos of her fantasy world. They constitute, rather, fragments from parallel worlds of experience and parallel texts inscribed in both individual (the artist's) and collective (the viewer's) memory. From this point of view, Porter's art is an art of memory, whose strategy of signification depends upon the activation of a suppressed system of cultural codes. Walter Benjamin's notion of the fragment, as the site of 'involuntary memory' where 'certain contents of the individual past combine with material of the collective past,' can serve to explain the logic and appeal of Porter's art" (Ramírez, "Illusive Fragments" 14).

6 At the time, Porter was part of the New York Graphic Workshop (1964–1970), an experimental collective she cofounded with Luis Camnitzer and José Guillermo Castillo that sought to reinvent printmaking. During those years, Noé, who happened to be Camnitzer's roommate, served as an informal mentor and intellectual interlocutor to the group. For more on the New York Graphic Workshop, see Pérez-Barreiro, Davila-Villa, McDaniel Tarver, and Blanton Museum of Art.

7 Noé wanted to translate that reality onto the canvas and spoke about "shattered vision" (visión quebrada) to describe the formal response to such chaos: "The key to a style of painting that is inspired by a shattered vision can be found in the struggle and the huge variety of elements that exist in a chaotic society. It no longer makes sense to speak of a single focal point, of only one optical center, or of just one strong line in a painting. Why not 48 or 64? The key to painting inspired by a shattered vision is right there, in the variety of lines flowing in divergent directions, chaotically invading the complex environment created by many opposing focal points" [Las claves de una pintura de visión quebrada están dadas por la lucha y multiplicidad de elementos tal como en una sociedad caótica. Ya no tiene sentido hablar de un punto de concentración visual, de un solo centro óptico y de una sola línea de fuerza en la obra. ¿Por qué no 48 o 64? Allí en la multiplicidad de líneas de fuerza en sentidos divergentes y opuestos y que entran a invadir el espacio real caóticamente, y en la multiplicidad de varios centros ópticos está la clave de una pintura de visión quebrada] (*Antiestética* 197).

8 In the interview with Ramírez, Porter claims, "for my work to be like when, for example, you are watching a movie, totally consumed by the illusion of the movie and suddenly, someone turns on the light. It bothers you, because you get distracted by the light and you can't give yourself completely to the movie. I think my work is like that, like watching a movie with the lights on" [que mi obra fuera como cuando uno, por ejemplo, está viendo una película, está totalmente entregado a la ilusión de la película y de golpe ahí prenden la luz y molesta, porque uno se distrae con la luz y no puede entregarse completamente a la película. Entonces yo pienso que mi obra es como si fuera eso, como si estuviera uno viendo una película con la luz prendida] (in Ramírez, "Entre espejos" 84).

9 For a history of cabinets of curiosities, see Mauriès. For a study about the origin of museums and their kinship with cabinets of curiosities, see Impey and MacGregor's edited volume and especially their introduction.

10 It is interesting to note that this organization appears, at first sight, as objective and scientific. However, in the cabinets, this objectivity quickly became improbable. Often, objects belonged to more than one category. The decision about where to display an object naturally conditioned its interpretation and determined the act of imagination that gave access to the promised alterity.

11 For an eloquent critique of this practice, see Fusco.

12 About this discussion as it manifests in filmmaking from Latin America, see Fornoff and Heffes.

Works Cited

Bazzano-Nelson, Florencia. *Liliana Porter and the Art of Simulation.* Ashgate, 2008.

Borges, Jorge Luis. *Other Inquisitions, 1937–1952.* Translated by Ruth L. C. Simms. University of Texas Press, 1964.

Fornoff, Carolyn, and Gisela Heffes, eds. *Pushing Past the Human in Latin American Cinema.* SUNY Press, 2021.

Fusco, Coco. "The Other History of Intercultural Performance." *TDR* 38, no. 1 (1994): 143–167.

Grinblatt, Julio. "270 Film Series, Julio Grinblatt on Liliana Porter, ISLAA." Video, 5:17, posted to YouTube, 2023. https://www.youtube.com/watch?v=YOW6omN0vYw.

Hobbes, Thomas. *Leviathan.* Penguin Classics, 2017.

Impey, O. R., and Arthur MacGregor. Introduction to *The Origins of Museums: The Cabinet of Curiosities in Sixteenth and Seventeenth-Century Europe,* edited by O. R. Impey and Arthur MacGregor, xvii–xx. Clarendon Press/Oxford University Press, 1985.

Macel, Christine. *Viva Arte Viva: Biennale Arte 2017.* Exhibition catalog. La Biennale di Venezia, 2017.

Mauriès, Patrick. *Cabinets of Curiosities.* Thames and Hudson, 2002.

Merewether, Charles. "Liliana Porter: The Romance and Ruins of Modernism." In *Liliana Porter: Fragments of the Journey,* 41–63. Exhibition catalog Bronx Museum of the Arts, 1992.

Noé, Luis Felipe. *Antiestética.* 2nd edition. De la Flor, 1988.

Pérez-Barreiro, Gabriel, Ursula Davila-Villa, Gina McDaniel Tarver, and Blanton Museum of Art, eds. *The New York Graphic Workshop, 1964–1970.* Blanton Museum of Art/University of Texas Press, 2009.

Porter, Liliana. *Liliana Porter: Fragments of the Journey.* Exhibition catalog. Bronx Museum of the Arts, 1992.

Porter, Liliana. *Liliana Porter: Obra gráfica 1964–1990.* Exhibition catalog. Instituto de Cultura Puertorriqueña/Comisión Puertorriqueña para la Celebración del Quinto Centenario del Descubrimiento de América y Puerto Rico, 1991.

Porter, Liliana. "Liliana Porter—*Wrinkle.*" Photo series. https://lilianaporter.com/pieces/229/assets/269.

Porter, Liliana, Inés Katzenstein, and Gregory Volk. *Liliana Porter: In Conversation with = En Conversación con Inés Katzenstein.* Fundación Cisneros/Colección Patricia Phelps de Cisneros, 2013.

Ramírez, Mari Carmen. "Entre espejos, sombras y viajeros: Liliana Porter y la poética del grabado." In *Liliana Porter: Obra gráfica 1964–1990,* 80–86. Exhibition catalog. Instituto de Cultura Puertorriqueña/Comisión Puertorriqueña para la Celebración del Quinto Centenario del Descubrimiento de América y Puerto Rico, 1991.

Ramírez, Mari Carmen. "Illusive Fragments: Liliana Porter's Art of Memory." In *Liliana Porter: Fragments of the Journey,* 13–39. Exhibition catalog. Bronx Museum of the Arts, 1992.

Speranza, Graciela. "A contratiempo." In *Liliana Porter: El hombre con el Hacha y otras situaciones breves,* 10–15. Fundación Eduardo F. Costantini, 2013.

II

Natural History and Ancestral Knowledge

4

Reactivating Richard Spruce's Amazonian Biocultural Archive

Luciana Martins

In his remarkable artwork *Broken Spectre* (2022) Richard Mosse provides a multiscalar vision of the Amazon as dystopia. His multimedia installation combines close-ups of forest insects among tiny plants shot in a dark studio with ultraviolet lenses, human-scale monochrome photographs of illegal miners and loggers, cowboys, and Yanomami Indigenous people trying to protect their land, culminating in astonishing brightly colored aerial photographs produced by drone-borne multispectral cameras and GIS-generated aerial maps showing the scale and extent of the forest destruction. Mosse explains that by using the latest surveying technologies, his aim is to explore their potential, subverting them to show "what the camera cannot see"; he displays the interdependent complexity of the Amazonian biome and its ruination to great effect.

Mosse's multimedia work gives visibility to a myriad of human and nonhuman agents, challenging the oft-invoked trope of wilderness in descriptions of Amazonia. In this sense, it bears some resemblance to the work of photographer and environmental activist Subhankar Banerjee, who produces colorful photographs of the Arctic to counterpoint the idea of the region as "flat, white, nothing there, frozen wasteland of snow and ice, barren wasteland" (in New Mexico PBS). Banerjee contends that "anywhere you ask people—in Argentina, in Africa, in India, what does it [the Arctic] look like? People will say, well, it's white; or the Amazon: it's green; or New Mexico, they'll say brown" (in New Mexico PBS). For Banerjee, the use of color to represent the Arctic is a political act, revealing the diversity of life in the region. His work draws attention to multispecies lifeworlds that are unfamiliar to the Western eye or perhaps made too familiar through TV nature documentaries that tend to obliterate Indigenous ancestral ways of coinhabiting these regions. In the Amazon, what the Western eye sees as wild is the result of dynamic human interaction with the tropical forest environments over millennia, a lengthy process that enabled humans and nonhumans to "live in common" (to use Gan and Tsing's expression, 103). As archaeologist Michael J. Heckenberger observes,

> There are few places on Earth where "nature" looms so large in the Western imagination as the Amazon. The earliest European explorers were awed by its vast natural resources, which today are coveted by developers and environmentalists alike. For the past few centuries, it was viewed as the setting, par excellence, of pristine nature and primitive tribes, the alter ego of Western civilization and built environment. In cultural terms, most people have a pretty good idea of what the average native Amazonian is like . . . naked, painted, and feather-bedecked bodies . . . living more or less "at one with nature." This image is deeply rooted in the Western imagination and appears time and again in feature films, coffee-table books, and magazines and on the Internet. (26–27)

The rhetoric surrounding the Amazon as wilderness and therefore an empty space to be conquered echoes the US ideology of Manifest Destiny. In Brazil, during President Getúlio Vargas's Estado Novo in the 1940s, the region played a key role in the country's push for geopolitical hegemony in South America supported by its "March to the West" public policy. In a speech in Manaus on October 10, 1940, Vargas made his intentions clear: "Nothing will deter us from this undertaking which is, in the twentieth century, the greatest task for civilized man: to conquer and dominate the valleys of the great equatorial torrents, transforming its blind force and extraordinary fertility into disciplined energy" (In Garfield 21–22).

What Tracy Devine Guzmán calls the "trinity of Brazilian modernity"—"technological progress, national security, and economic development"—would inform national development policies in the region throughout the period of Brazil's military dictatorship (1964–1985), culminating in the hyperbolic, dystopic version embodied by former president Jair Bolsonaro (2019–2022) (Devine Guzmán 109).

In August 2023, a two-day meeting of the eight-member Amazon Cooperation Treaty Organization of Bolivia, Brazil, Colombia, Ecuador, Guyana, Peru, Suriname, and Venezuela, was convened by Brazilian President Luiz Inácio Lula da Silva, who told the audience, in sharp contrast to Bolsonaro's position, "The rainforest is neither a void that needs occupying nor a treasure trove to be looted. It is a flowerbed of possibilities that must be cultivated" (in Phillips). The resulting joint communiqué, the Belém Declaration (Brazil, "Nota à imprensa"), pledges to strengthen regional cooperation, acknowledging the land rights of Indigenous peoples and their role in biodiversity conservation while, as many environmentalists point out, failing to address openly the main causes of deforestation and commit to zero deforestation by 2025 (Gonçalves; Varshney). The Brazilian president made it clear that international support from the Global North is vital for this initiative to succeed, urging "rich countries which have already destroyed their forests [to] take responsibility for funding our development" (in Phillips).

This prologue on the trope of Amazonian wilderness and its associated deforestation might initially seem unrelated to this book's topic: What does it have to do with a renewed interest in natural history collections in Latin America? The Indigenous activist, researcher, and artist Glicéria (Célia) Tupinambá provides one answer. Since 2004, she has been involved in the struggle for the recognition of Indigenous territorial rights in southern Bahia state in Brazil and the subsequent rewilding of deforested areas (Tupinambá, "An Indigenous Woman" 15–16). Her emotional encounter with an ancient Tupinambá feather mantle in a European museum in 2018 triggered a personal journey to relearn how to make it, prompting her to mobilize a cosmopolitical set of relations between human and nonhuman entities. This included listening to kin, friends, allies, elders, and the mantle itself, paying attention to the signs of ancestors and the *encantados* in her dreams, as well as being alert to the gestures of forest plants, bees, and birds.[1] "The cloak can be made only if the territory is alive," she insightfully affirms (Tupinambá, "The Cloak"). For Tupinambá, who has since visited various museums in Europe, these institutions should be "a space of dialogue" and "not the space of extinguished memories," she contends, "but the place of a new story, with another shade of history" ("An Indigenous Woman" 22). As Tupinambá puts it,

> We have landmarks, and our landmarks are in museums, and they continue to look after us. And people make sure to look after them. There are 11 mantles in Europe.[2] I see this as a brilliant idea on the part of the Tupinambá. Having artefacts locked up so that they then become landmarks of occupation and territorial reach, bearing witness to the fact that the Tupinambá *also occupied Europe.* ("An Indigenous Woman" 21, emphasis in the original)

It is by retracing these landmarks, "footprints," as Tupinambá refers to them in a roundtable, that the memories of the mantles can be reactivated by the Tupinambá people (in Pitt Rivers Museum). In contrast to the mantles locked up in museums, Tupinambá's new mantles are incorporated into the life of the community, in movement. Her case is an inspiration, showing how museum collections in the Global North can support the reactivation of ancestral knowledge related to biodiversity in the Global South. Following her path, I explore the ways in which the Amazonian biocultural archive in European institutions and elsewhere has the potential to speak to pressing questions around Indigenous socioenvironmental heritage and its sustainability in the twenty-first century.

Seen from the perspective of Western science, the biocultural archive created in the context of European imperial and colonial expansion in the tropical world since the sixteenth century—composed of biological specimens, cultural objects, images, and documents—today comprises an often-neglected treasure trove of ecological and sociocultural data (Salick, Konchar, and Nesbit 1). Ethnographic artifacts, plant and animal specimens, manuscript notes, drawings, letters, and journals collected in museums and herbaria hold valuable information about the natural, cultural, and social worlds through which European naturalists traveled. Rather than reiterating heroic collection narratives, it is possible to engage with these collections to recover social, cultural, and ecological histories that too often remain dormant in the archives. In pursuit of exotic plants, animals, and artifacts, European explorers typically "walked together" in the field with Indigenous peoples, benefiting from their expert knowledge of the local environment, plants, and animals (Davis 184). From their archives, it is possible to recover aspects of past ecological relations between humans and other species, enabling the environment to assume a kind of agency. In this context, the biocultural archive presents a fundamental resource for Indigenous peoples seeking to reestablish their connection with entities from the past and to reactivate ways of understanding and being in the world (Martins 36–38). The heritage reconstituted by research on families of objects, specimens, and archives effectively materializes Indigenous relations to human, nonhuman, and other-than-human entities, creating new tools for addressing present and future socioenvironmental challenges (Augustat 239–240; Daly and Shepard 17).

This chapter results from reflections on a research program initiated in 2015 dedicated to reactivating the biocultural archive created by British botanist Richard Spruce, who spent fifteen years (1849–1864) traveling in the Amazon and the Andes. Involving researchers at Birkbeck, University of London, and the Royal Botanic Gardens, Kew, the program has developed in partnership with the Botanical Garden of Rio de Janeiro, the Socio-Environmental Institute, and the Federation of the Indigenous Organizations of the Rio Negro, among other institutions in Brazil (Fonseca-Kruel et al. 218–225).[3] Spruce's unique collections, today housed mainly at Kew Gardens and the British Museum in London, incorporate various Indigenous types of artifacts, including plant-based objects, samples of useful plant products, detailed archival notes on the use of plants by local inhabitants, accompanying herbarium voucher collections, vocabularies, and drawings. In this chapter, I focus specifically on the example of the caraipé bark used to make fireproof pottery. By connecting

Spruce's materials with those of other traveling naturalists, I hope to shed light on the relations between humans and the nonhuman world that remain latent in the European biocultural archive. By revisiting this archive with renewed attention, we can contribute to environmental conversations related to the diversity of life in the Amazon region relevant to the challenges we face today.

Documenting Fireproof Pottery

In addition to funding his travels through the sale of his collections of dried specimens, seeds, and living plants (Hooker, "Mr. Spruce's Intended Voyage"), Spruce also made arrangements to send William Hooker, director of the Royal Botanic Gardens, Kew, "botanical objects" for Kew's newly created Museum of Economic Botany (Hooker, "Botanical Objects" 169). Opened in 1847, the botany museum was "designed to bring together in one spot and to exhibit those interesting vegetable products from all parts of the world" (Hooker, "Botany" 405). As Hooker enumerates in the "Botany" section of the Admiralty's *Manual of Scientific Inquiry*, published in 1849, objects to be collected for the museum included fruits and seeds; flowers; entire plants, or parts of them; trunks of trees; woods; gums and resins; dye-stuffs of various kinds; medicinal substances; and general products of vegetables ("Botany" 405–408).

Spruce was a diligent collector, corresponding with Hooker regularly to inform him of his activities. Hooker, for his part, published Spruce's letters in *Hooker's Journal of Botany and Kew Garden Miscellany*, keeping the subscribers informed about the progress of Spruce's collecting expedition while also publicizing his endeavors.[4] Within nearly three months of his arrival in Belém, on October 7, 1849, Spruce sent Hooker a letter that includes a list of "vegetable curiosities" that he had amassed for the museum, "which fill two cases" (Spruce, "Botanical Excursion" 65). In this letter, Spruce first described the making of pottery:

> My principal excursion from Caripi was to an Indian settlement in the heart of the forest, about five miles from Mr Campbell's house,[5] where the manufacture of the *Caraipé*, or fire-proof pottery (a branch of art still confined to the Indians), is carried on. Our journey was certainly an extraordinary one: along hunters' tracks, which none but an Indian could have found, over fallen trees, and occasionally across an igaripé [*igarapé*, artificial canal], the only means of crossing by a single trunk of a tree, which I and my companion were glad to traverse *à cheval* [on horseback], much to the amusement of our guides, who tripped across with their bare feet in security. After witnessing the process of making

> pottery, an old Indian accompanied us about two miles further into the forest, to see the Caraipé-tree growing, and, if possible, to procure specimens of it, which we succeeded in doing, though unfortunately neither flower nor fruit were visible. ("Botanical Excursion" 66)

In this passage, Spruce draws attention to several skills of the Indigenous people, including their art in making pottery, ability in wayfaring, sense of humor, and expertise in identifying specimens in the intricate forest. These skills would be vital to the success of Spruce's botanizing enterprise, especially because he was already regretting bringing with him from England Robert King, "a young man who has agreed to brave the wilds of the Amazon as my companion and assistant" (Spruce, *Notes of a Botanist* 1:1). Spruce explains in a letter to Hooker that his companion was "of rather plethoric habit, & his activity seems to have forsaken him." He then lists a few potential alternate assistants, including a Black man who had been employed by the Austrian naturalist Johann Natterer during his Brazilian expedition and two other Frenchmen.[6] After King left the expedition in April 1850, Spruce found intermittent assistance from Indigenous and local people as he traveled along the Amazon region (*Notes of a Botanist* 1:210).

The vegetable curiosities that Spruce sent to the Kew Museum from this first excursion included leaves and bark of the pottery tree collected in August 1849. In the entry for these items on the accompanying list, Spruce details the tree's appearance and measurements as "exceedingly straight, slender, and lofty, attaining a height of 100 ft. before it sends forth branch, and with a diameter at the basis not exceeding 12–15 inches"; the wood was "so hard that our tools would not enter it" ("Botanical Excursion" 73). He goes on to explain that to make the caraipé pottery, "the purest clay is preferred" and "is procured from the beds of the rivers and igarapés." He added that the utensils he sent, including two *panelas* (pans) "used for heating milk, boiling eggs, and similar purposes" and a chafing dish (figure 4.1), were made by an Indian woman who lived on the Igaripé Castanhal at Tanau and consisted of "nearly equal portions of clay and the powdered bark of the Caraipé" ("Botanical Excursion" 73).

Three months later, near the Trombetas River in a place now called Praia de Caipurú, about 180 miles west of Santarém, Spruce realized that he misunderstood the explanation given of the process of making pottery in Caripi. He learned from the sister of his host at the Santa Cruz farm, Dona Cesária, that the caraipé "bark *was first burnt,* and the ashes then mixed with clay" ("Journal of an Excursion" 276). This process increased the tenacity of the pottery. He obtained a quantity of the ashes from her and sent them to the museum with notes on the geographical spread of the caraipé tree and further uses of the pottery made with its bark.

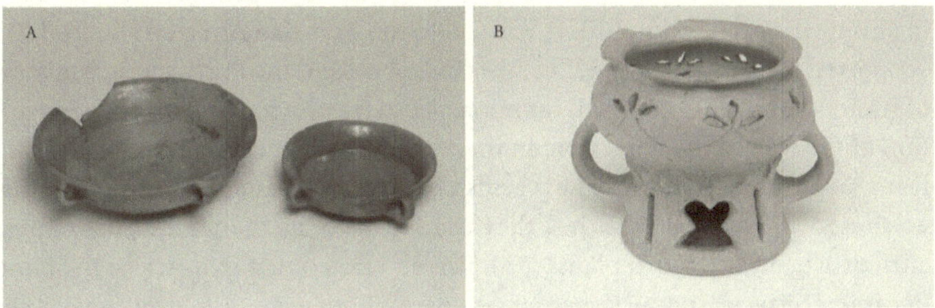

Figure 4.1. Fireproof pottery collected by Richard Spruce in the Brazilian state of Pará in 1849. (*A*) Pans of fine clay and powdered tree bark that has been burned (EBC 37794); (*B*) chafing dish made of clay and bark (EBC 37798) © RBG Kew. CC-BY 4.0.

Spruce traced the use of caraipé along all the tributaries of the Amazon River, "along the eastern roots of the Andes of Peru and Ecuador, or in the provinces of Maynas and Canelos" (*Notes of a Botanist* 1:14). He details the tree-climbing skills of the people there as well as their ability to identify the caraipé's quality: "The Indians test Caraipé by burning it; and if, when burnt, it cannot be broken by the fingers, but requires the use of a mortar, it is considered good" ("Journal of a Voyage" 42). He notes that "the best Caraipé is afforded by species growing in rich, dryish soil; those of the low sandy forests, and of the gapó [igapó, flooded blackwater forest], containing only a small proportion of silex in the bark" ("Journal of a Voyage" 42). Spruce sent the leaves and later flowers and fruits of the caraipé tree to George Bentham, who identified it as a species of *Licania* of the Chrysobalanaceae order. The herbarium specimens of caraipé sent by Spruce thus confirm his account, the early ones with just leaves and the later ones with flowers and fruits, attesting to his different collecting locations in the Amazon.

Spruce noted that the caraipé was already a celebrated tree, having been first mentioned in the botanical literature by the French pharmacist and botanist Fusée Aublet in his account of the flora of French Guiana published in 1775 (Spruce, "Botanical Excursion" 73).[7] Accompanying an illustration of the trees' leaves and branches, Aublet thus described it: "This tree is called 'Caraipé' by the Garipons. They use the ashes of the bark mixed with a clayey soil to make their pottery. The Creoles call this species 'Manche-hache' ('Axe handle') because the wood is one of the best for making handles of axes, hatches and other cutting tools" (in Ploktin, Boon, and Allison 3).[8]

More than ten years later, in 1786, the Portuguese naturalist Alexandre Rodrigues Ferreira sent a note from the village of Barcelos, on the bank of the

Rio Negro, to accompany the ceramic pieces that he shipped to Portugal during his nine-year "Philosophical Voyage" to the Brazilian Amazon (figure 4.2), supported by the Academy of Sciences in Lisbon and the Portuguese Ministry of Business and Ultramarine Dominions. Ferreira provided a detailed description of the way Indigenous women made pottery from caraipé bark, the resin from "intaiúca" that they used to varnish the pieces, the process of smoking the upside-down pans, and the dyes they used to decorate the pieces from tanha, curi, urucu, and carajuru plants (75). Ferreira noted that in a lack of the bark of caraipé, crushed pottery shards were used; if not even those were available, calcinated turtle shields would be mixed with river clay for toughening ceramics (75). It is worth noting that the "portable library" Ferreira took with him to Brazil included Aublet's *Histoire the plantes de la Guiane Française* (Safier 117). As Neil Safier remarks, reading was for Ferreira a key activity that influenced what to observe in situ "through the cross-fertilization of ocular observation and bibliographical consultation" (124). Ferreira's own writing, however, would not be incorporated into the baggage of subsequent nineteenth-century travelers to the Amazon region, since his scattered collection remained unpublished until the late twentieth century.[9]

The process of using caraipé for making pottery was also documented by British naturalists Alfred Russel Wallace and Henry Walter Bates, who arrived together in the Amazon in 1848. Wallace mentioned asking his Black male servant "Isidora" (Isidoro) to show them where they could "obtain specimens of a tree called Caripé, the bark of which is used in the manufacture of the pottery of the country" (22).[10] A free man who had worked for Englishmen before "as cook and servant-of-all-work" (Bates 1:10), Isidoro apparently displayed great knowledge of the local trees and their uses, leading Wallace and Bates to a young caraipé tree "with neither fruit nor flowers"; they had therefore to content themselves "with specimens of the wood and bark only" (Wallace 23). In his own published travel account, Bates also makes reference to the use of caraipé for pottery-making: "To enable the vessels to stand the fire, the bark of a certain tree, called Caraipé, is burnt and mixed with the clay, which gives tenacity to the ware. Caraipé is an article of commerce, being sold, packed in baskets, at the shops in most of the towns" (2:40). Bates's entomological eyes and ears, however, were more attuned to the deposits of the stiff white clay on the riverbank than to the tree: "The shallow pits, excavated in the marly soil at Mahicá [east of Santarém], were very attractive to many kinds of mason bees and wasps, who make use of the clay to build their nests with" (2:40).

Bates then describes in detail the making of the nests by a large yellow and black wasp, the *Pelopaeus fistularis* (*Sceliphron fistularium*), including the "loud hum" it made when arriving at the pit, collecting the clay "in little round pellets,

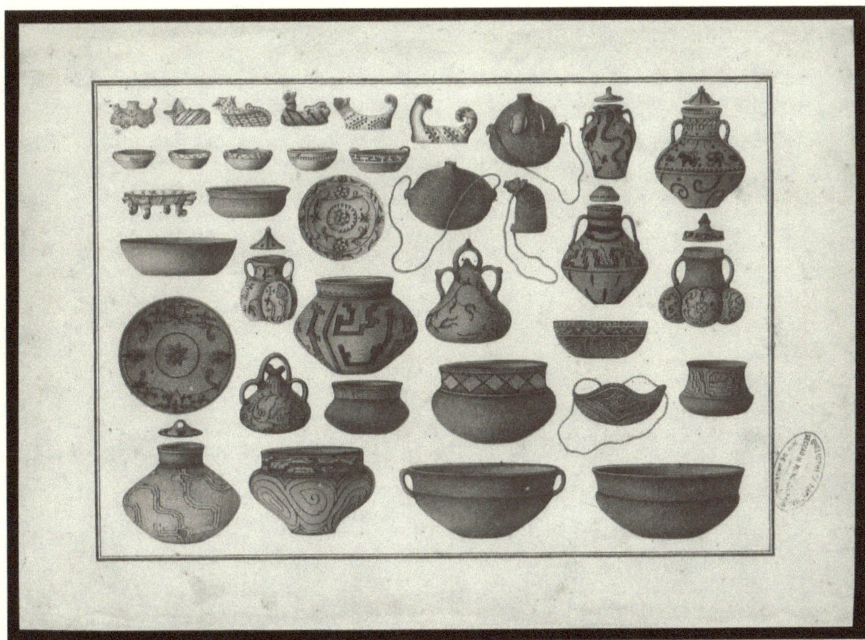

Figure 4.2. Drawings of Indigenous domestic utensils, Alexandre Rodrigues Ferreira expedition, n.d. Ink on paper, 27.5 cm × 19.5 cm. Acervo da Biblioteca Nacional–Brasil.

rolling them into a convenient shape in its mandibles" (2:40), and the "cheerful busy hum when it alighted and began to work," taking about a week to build a cocoon in sunny weather (2:41). He observed similar habits in three kinds of *Trypoxilon* (*T. albitarse, T. aurifrons,* and an unnamed species) that were also building wasps. But it was the work of a social bee, *Melipona fasciculata,* that he found most intriguing (figure 4.3). Bates observed how the bees used their hide shanks adapted for the collection of pollen to hold the clay pellets that they transported to build up a wall in the construction of "their combs in any suitable crevice in trunks of trees or perpendicular banks" (2:44). Despite having no sting, he found, "they bite furiously when their colonies are disturbed" (2:45).

There is no evidence that either wasps or bees mix caraipé bark with clay to make their nests, but there is something about their humming and intensity of the work of collecting clay and molding it that resemble the work of the Indigenous ceramists, normally a female activity.[11] Nesting and making artifacts for processing and making food are caring activities that enable the reproduction of life in the forest, establishing a connection with Earth's embodied materiality. They require specialized knowledge, understanding, and intimacy with the

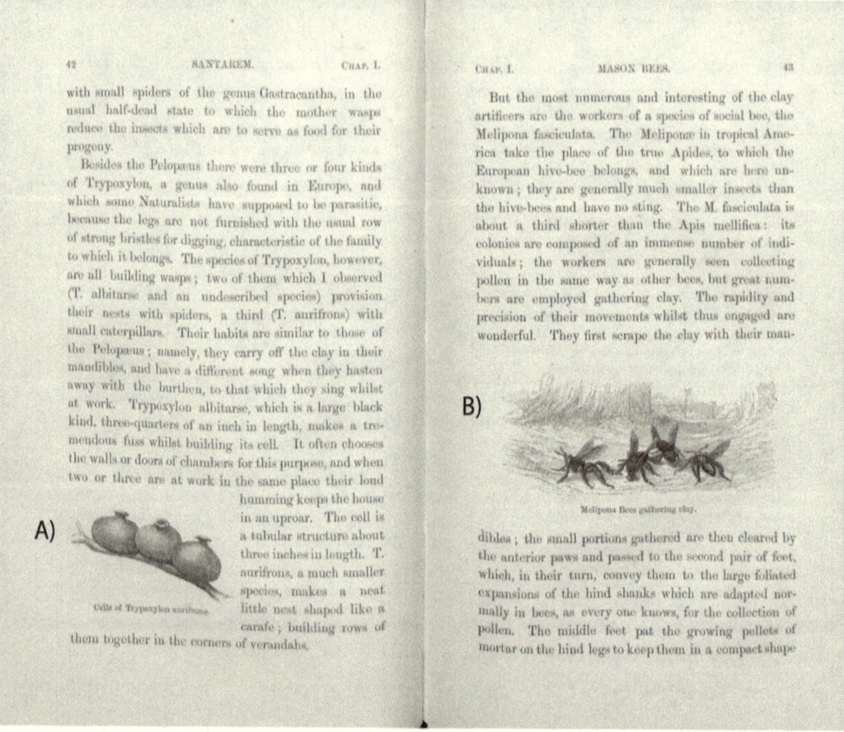

42 SANTAREM. CHAP. I.

with small spiders of the genus Gastracantha, in the usual half-dead state to which the mother wasps reduce the insects which are to serve as food for their progeny.

Besides the Pelopæus there were three or four kinds of Trypoxylon, a genus also found in Europe, and which some Naturalists have supposed to be parasitic, because the legs are not furnished with the usual row of strong bristles for digging, characteristic of the family to which it belongs. The species of Trypoxylon, however, are all building wasps; two of them which I observed (T. albitarse and an undescribed species) provision their nests with spiders, a third (T. aurifrons) with small caterpillars. Their habits are similar to those of the Pelopæus; namely, they carry off the clay in their mandibles, and have a different song when they hasten away with the burthen, to that which they sing whilst at work. Trypoxylon albitarse, which is a large black kind, three-quarters of an inch in length, makes a tremendous fuss whilst building its cell. It often chooses the walls or doors of chambers for this purpose, and when two or three are at work in the same place their loud humming keeps the house in an uproar. The cell is a tubular structure about three inches in length. T. aurifrons, a much smaller species, makes a neat little nest shaped like a carafe; building rows of them together in the corners of verandahs.

Cells of Trypoxylon aurifrons.

CHAP. I. MASON BEES. 43

But the most numerous and interesting of the clay artificers are the workers of a species of social bee, the Melipona fasciculata. The Meliponæ in tropical America take the place of the true Apides, to which the European hive-bee belongs, and which are here unknown; they are generally much smaller insects than the hive-bees and have no sting. The M. fasciculata is about a third shorter than the Apis mellifica: its colonies are composed of an immense number of individuals; the workers are generally seen collecting pollen in the same way as other bees, but great numbers are employed gathering clay. The rapidity and precision of their movements whilst thus engaged are wonderful. They first scrape the clay with their mandibles; the small portions gathered are then cleared by the anterior paws and passed to the second pair of feet, which, in their turn, convey them to the large foliated expansions of the hind shanks which are adapted normally in bees, as every one knows, for the collection of pollen. The middle feet pat the growing pellets of mortar on the hind legs to keep them in a compact shape

Melipona Bees gathering clay.

Figure 4.3. Illustrations of wasp nests and bees gathering clay, in Henry Walter Bates, *The Naturalist on the River Amazons*, vol. 1, 1863. (*A*) "Cells of Trypoxylon aurifrons"; (*B*) "Melipona bees gathering clay." Library of Congress, Rare Book and Special Collections Division, https://www.loc.gov/item/49032931/.

elements. As Deborah Bird Rose asserts, "The Western distinction between organic and inorganic matter is irrelevant within connectivities that permeate all matter" (33).

In their different registers, the biocultural archive of these naturalists provides glimpses of particular socioenvironmental processes in specific places that call attention to the biodiversity of the Amazon. In this view, instead of "wilderness," a homogeneous green hell or bountiful rainforest, the region becomes a meshwork of human and nonhuman activities articulated in particular knots (Ingold 70). Although caraipé pottery-making took place across several Indigenous communities in Amazonia and beyond and to a certain extent still does, attention to this rich biocultural archive brings to light subtle differences and their commonalities. The pots collected in Santarém by Spruce are quite different from those collected by Ferreira in Barcelos. And there are several species of *Licania,* with different qualities for pottery-making. Spruce reported during a collecting expedition near Óbidos along the Trombetas River that he identified another kind of caraipé, *Licania turiuva,* a synonym of *Licania octandra* (Santos-Fonseca, Coelho-Ferreira, and Fonseca-Kruel 225). As Spruce describes the tree,

> [It] is called Caraipé das agoas [águas], and its calcined bark is used in the fabrication of pottery, in default of better, for it contains a very small proportion of silex. It is commonly remarked among the Indians that the products of trees of the gapó [igapó]—whether bark, timber, fruits, or resins—are inferior to those of other trees, their congeners, growing on terra firme or beyond the reach of floods. (*Notes of a Botanist* 1:87)

Among the accounts of the use of caraipé in Amazon pottery-making described here, Spruce's is the only one that mentions silica as the element that increases the resistance of pottery, even reporting that in "the best sorts, the silex can be seen by the naked eye, filling up the vessels of the bark" ("Journal of a Voyage" 42). A note on the "Pottery Trees" published in a Royal Botanic Garden bulletin in 1903 reproduces a letter by Spruce dated June 2, 1880, in which he replies to a query on a piece of pottery at Kew Museum: "If I recollect rightly, the tree which afforded the best Caraipé on the Naupés [Uaupés River] was one I sent specimens of under the name *Licania crassivenia*" (Royal Botanic Garden 26). Collected near the Uaupés River in November 1852, the label on the type specimen in the Kew herbarium states that Spruce's taxonomy was published in Martius's *Flora brasiliensis* (figure 4.4). In the same note, there is a mention of the bark being sent to the mineralogist Walter Flight for him to determine the amount of silica in its ash. Flight reported that after pounding the bark into a mortar and drying it at 110 degrees Centigrade, he found that 18.19 percent of the outer bark was silica, with the remaining composed "entirely of lime and potash sulfate" (Royal Botanic Garden 26–27).

The role of silicon in ecological processes in tropical forests has only recently garnered attention from environmental scientists, who suggest that its functions include alleviation from the stress in plants caused by drought, salinity, and extreme temperatures as well as the damage caused by living organisms. Jörg Schaller and colleagues report that among other chemical processes, silicon is an important factor in the global carbon cycle (Schaller et al. 162) and that it "is also likely to play an important role in tropical forests for species performance and distribution, for community composition and for responses to climate change" (171–172), indicating the possibility of examining historical data for the understanding of its evolutionary ecology. In their materiality, Spruce's leaf, bark, and ash samples might have a relevant role in such investigations.

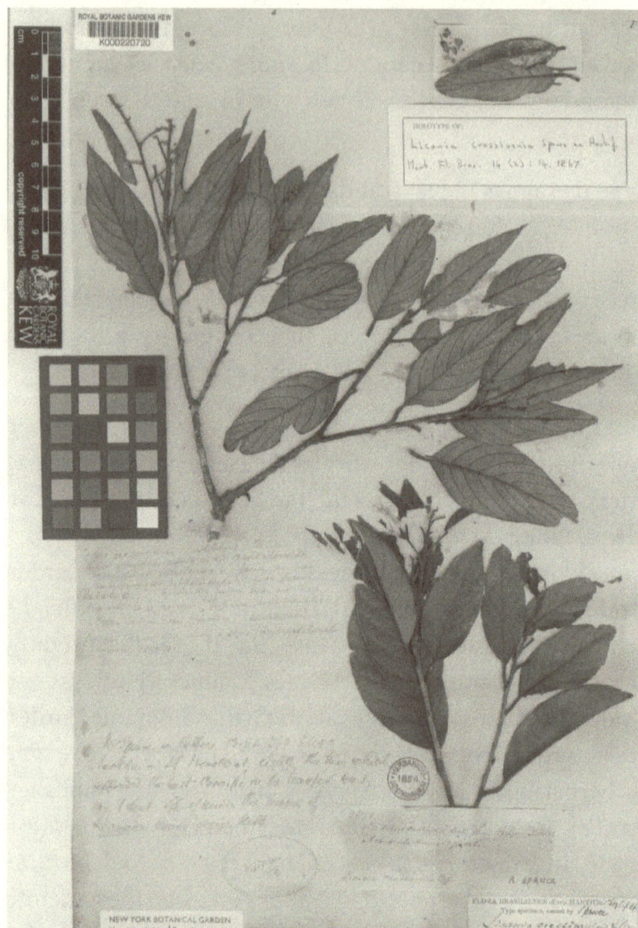

Figure 4.4. *Licania crassivenia* Spruce ex Hook.f. specimens collected by Richard Spruce in the Uaupés in November 1853 (No. 2678; barcode K000220720) © RBG Kew. CC-BY 4.0.

Archaeology is another area of research in which there is significant potential use for such biocultural collections. While the materials produced about the pottery tree during the expeditions of eighteenth- and nineteenth-century naturalists have been dismissed by one archaeologist as "merely mentions of the uses of certain plants" [sendo meramente menções do uso de determinadas plantas] (Hepp 34),[12] attention to the objects collected as well as to the accounts and notes related to them provide a rich source for understanding changing uses of caraipé over time. Furthermore, they also reveal how widespread the knowledge of its use for pottery-making was, reaching riverine peoples well beyond the Indigenous communities.

The biocultural archive of caraipé pottery-making in the Amazon includes dried leaves, flowers, and fruits; pots, pans, and vases; notes and publications on peoples, trees, bees, wasps, clay, and riverbanks; and illustrations associated with all of these. To effectively reactivate the knowledge dormant in these collections, we need to walk together in a more symmetrical way than the travelers did, retracing the itineraries of the collected items, museum curators, scientists, social scientists, academics, and practitioners in the arts and humanities side by side with Indigenous researchers, practitioners, and artists. This reassemblage will make us more aware of the whispers between human and nonhuman entities that have remained muted in the archives for too long and will enable the recovery of Indigenous biocultural knowledge that has been lost over the last centuries.

Resources of Hope: Reactivating Biocultural Collections and Lost Connections

Resources of Hope is the title of a volume of posthumously collected essays by cultural theorist Raymond Williams edited by Robin Gable. Williams envisioned a powerful alternative to what today we call neoliberalism: "a radically new kind of politics coalescing the disarmament, environment and feminist movements" (O'Brien 739). Delineating the elements of an "ecologically conscious socialism," Williams imagined "the outline of political bearings which can be related to material realities in ways that give us practical hope for a shared future" (Williams 225). Williams died in 1988, not living to witness the magnitude of planetary damage that we face today.

Thirty years later, Elaine Gan and colleagues argue in the introduction to their collection *Arts of Living in a Damaged Planet,*

> "Anthropocene" is the proposed name for a geologic epoch in which humans have become the major force determining the continuing liveability of the earth. The word tells a big story: living arrangements that took millions of years to put in place are being undone in the blink of an eye. The hubris of conquerors and corporations make it uncertain what we can bequeath to our next generations, human and not human. The enormity of our dilemma leaves scientists, writers, artists, and scholars in shock. How can we best use our research to stem the tide of ruination? (1)

Richard Mosse's artwork *Broken Spectre,* which opens this chapter, surely forms part of this quest to curb the tide of ruination by forcing us to pay attention to the devastation of the Amazon rainforest currently underway. The Amazon is a fitting example of an anthropogenic landscape that Gan and colleagues depict as "haunted by imagined futures" (2) that go in tandem with modernity's "complex of symbolic and material projects for separating 'nature' and 'culture'" (7); yet "living landscapes are imbued with earlier tracks and traces" (6). Researching the biocultural archive in European museums provides one way of recovering such tracks and traces. It enables the telling of partial, forgotten histories neglected in the past by the narratives of progress, which always look forward. Similar to the *Melipona fasciculata* bees that Bates encountered, the planet is reacting furiously to having its landscapes destroyed. As we are urged to become more aware of multispecies interdependence and the need to cross sciences, arts, and humanities, these collections and the connections they afford acquire new meaning. They become resources of hope.

In the wake of the devastating fire at the National Museum in Rio de Janeiro in September 2018, in which an irreplaceable collection of Indigenous artifacts including those from the Amazon was lost, its curators are rethinking the role of the museum in the twenty-first century. Instead of remaining a keeper of relics, the museum is assuming the role of mediator in the politics of memory, focusing on the act of collecting itself (J. Oliveira 19). This entails its articulation with Indigenous organizations, intellectuals, and students to inform how they want to be represented, respecting their particular protocols and political strategies, as well as the establishment of partnerships with museums and other collecting institutions in the Global North for sharing their collections digitally and making them available to the Brazilian public, especially the Indigenous communities (J. Oliveira 19–20).

Defying the geopolitical and economic imperatives that informed Brazilian policies in the Amazon from the late nineteenth century to the present, the National Museum curators' initiative indicates that in order to tackle current socioenvironmental challenges, it is important to slow down. The two strategies outlined above require time, listening to. They point toward the creation of a "holding environment," one that depends upon care, protection, and trust (Berg and Seeber 15). Above all, it is an environment that gives time for ideas to be discussed, building understanding, and the museum to be a space for dialogue, as Tupinambá suggests ("An Indigenous Woman" 22).[13]

In this chapter I draw upon the work of a multidisciplinary team engaged in a research program on Spruce's biocultural collections. The focus on the caraipé was prompted in particular because of the interest that the Indigenous participants showed in learning more about the plant and its uses as well as on the pottery samples housed at Kew and elsewhere. By reactivating their ancestral biocultural knowledge in partnership with botanists, they aim to improve their skills in making pottery while enhancing sustainable environmental practices. Alongside this work on Spruce's biocultural collections and in dialogue with them, the artist Lindsay Sekulowicz has produced delicate unfired vessels made with clay dug from riverbeds. She uses her artistic imagination to activate these collections, expanding their use "beyond the science of botany and taxonomy into new ways of knowing the world of plants and peoples" (Martins and Sekulowicz). Although the reflections in this chapter are mine, they evolved through dialogue with project partners over a decade.[14]

This research program requires long-term commitment, learning to listen, being open and flexible to enable the codevelopment of research goals, being prepared to face the challenges of trans- and interdisciplinary research, working with multiple institutions with different aims, having to navigate different geographical scales—the local, regional, national, transnational, and global; observing ethical norms and protocols related to research with human participants, genetic heritage, and Indigenous knowledges through consultation and negotiation; and providing adequate time and resources for Indigenous people to engage directly with the collections, to recover their memories, bringing them back to life, as Tupinambá proposes ("An Indigenous Woman" 22). Above all, following Tim Ingold, it means recovering the sense of astonishment that reconnecting histories, objects, peoples, and ecologies bear on the understanding of anthropogenic landscapes and their relevance for imagining alternative futures.

Notes

1 In Tupinambá cosmology, *encantados* are "the Enchanted Ones." The mantles "are perceived as mediators between the Indigenous peoples and the Enchanted Ones, allowing for a particular spiritual connection" (Tupinambá, "An Indigenous Woman" 23).

2 These mantles are held in European museums, including the Musée du quai Branly–Jacques Chirac in Paris, the Museum der Kulturen in Basel, the Musées Royaux d'Art et d'Histoire in Brussels, the "Museum Septalianum," Biblioteca Ambrosiana di Milano in Milan, and the Nationalmuseet Etnografisk in Copenhagen (Tupinambá, "An Indigenous Woman" 23). In July 2024, the Nationalmuseet returned one of its five cloaks to the National Museum in Rio de Janeiro, Brazil (Abdala; Buono 14; Weinholt-Ludvigsen).

3 For more details and online publications on the collaborative project, visit the website Digital Repatriation of Biocultural Collections, https://biocultural.wpengine.com/.

4 The botanist George Bentham named and distributed Spruce's dried collections to subscribers "at the price of £2 the 100 species" (Hooker, "Mr. Spruce's Voyage to Pará" 347).

5 Caripi was an estate of Archibald Campbell, a Scottish-born merchant based in Belém, on the island of Carnapijó.

6 Spruce's letter to Hooker of 3 August 1849 is archived with the locator WCP4899 in the Alfred Russel Wallace Correspondence Project, Epsilon digital archive, Cambridge University, and is available at https://epsilon.ac.uk/view/wallace/letters/WCP4899.

7 Aublet is considered a "founding father" of neotropical botany (Plotkin, Boon, and Allison 1).

8 Mark Plotkin, Brian Boon, and Malorye Allison note that the "Garipons disappeared from French Guiana by the turn of the century but survive in Brazil where they are known as the Karipuna" (2).

9 Ana Maria de Moraes Belluzzo observes that the "first partial edition of the Philosophical Voyage's records would only be published in 1970, by Editora Brunner in São Paulo . . . obtained through a despicable act of piracy" (65).

10 The servant's name was actually Isidoro. Spruce commented in a letter to Daniel Hanbury in 1865 discussing Wallace's book that "Wallace is a poor linguist, & makes frequent mistakes in his Portuguese, especially in the genders."

11 For a detailed description of pottery-making by the Baniwa and Tukano women in the upper Rio Negro today, see, respectively, Thiago da Costa Oliveira's *Cerâmica baniwa* and Juliana Lins's *Cerâmica tukano.*

12 Unless otherwise indicated, all English translations are my own.

13 Despite the National Museum staff's good intentions, on the occasion of the return of the Tupinambá mantle from Denmark in July 2024, Indigenous leaders were frustrated because they were prevented by the National Museum, due to conservation measures, from "performing the necessary rituals to receive the relic, which is considered sacred" (Rogero). This incident indicates the difficulties in reconciling Indigenous and scientific demands.

14 I am indebted to insightful exchanges with the program collaborators Viviane da Fonseca-Kruel, Mark Nesbitt, William Milliken, Lindsay Sekulowicz, Aloisio Cabalzar, Andrea Scholz, Laura Osorio Sunnucks, and Cynthia Sothers and Indigenous researchers Lucas Alves Bastos, Oscarina Caldas Azevedo, Guilherme Tenório, Tarcísio Barreto, Vilmar Azevedo, Rosivaldo Miranda, and above all, the late Dagoberto Lima Azevedo, whose generosity and commitment to recovering Indigenous socioenvironmental knowledge in the upper Rio Negro were remarkable.

Works Cited

Abdala, Vitor. "Museu Nacional confirma retorno de Manto Tupinambá ao Brasil: Peça foi para a Dinamarca há 355 anos, em 1689." *AgênciaBrasil,* 11 July 2024. https://agenciabrasil.ebc.com.br/geral/noticia/2024-07/museu-nacional-confirma-retorno-de-manto-tupinamba-ao-brasil.

Augustat, Claudia. "Conclusion, a Play in the Field of Words: From Material Culture to/and Cultural Heritage." *Indiana* 37, no. 22 (2020): 237–245.

Bates, Henry Walter. *The Naturalist on the River Amazons: A Record of Adventures, Habits of Animals, Sketches of Brazilian and Indian Life, and Aspects of Nature under the Equator during Eleven Years of Travel.* 2 volumes. John Murray, 1863.

Belluzzo, Ana Maria de Moraes. *A Place in the Universe: The Voyager's Brazil.* Volume 2. Translated by H. Sabrina Gledhill. Metalivros, 1995.

Berg, Maggie, and Barbara K. Seeber. *The Slow Professor: Challenging the Culture of Speed in the Academy.* University of Toronto Press, 2016.

Brazil, Ministério das Relações Exteriores. "Nota à imprensa nº 331: Declaração presidencial por ocasião da Cúpula da Amazônia—IV Reunião de Presidentes dos Estados Partes no Tratado de Cooperação Amazônica." Press release, 8 August 2023. https://www.gov.br/mre/pt-br/canais_atendimento/imprensa/notas-a-imprensa/declaracao-presidencial-por-ocasiao-da-cupula-da-amazonia-2013-iv-reuniao-de-presidentes-dos-estados-partes-no-tratado-de-cooperacao-amazonica.

Buono, Amy. "Seu tesouro são penas de pássaro: Arte plumária tupinambá e a imagem da América." Translated by Patricia D. Meneses. *Figura: Studies on the Classical Tradition* 6, no. 2 (2018): 13–29.

New Mexico PBS. "¡Colores!: Photographer Subhankar Banerjee." Video, 8:03. Posted to YouTube, 5 December 2016. https://www.youtube.com/watch?v=DFvJ79NVofI.

Daly, Lewis, and Glenn Shepard Jr. "Magic Darts and Messenger Molecules: Toward a Phytoethnography of Indigenous Amazonia." *Anthropology Today* 35, no. 2 (April 2019): 13–17.

Davis, Michael. "Walking Together into Knowledge: Aboriginal/European Collaborative Environmental Encounters in Australia's North-east, 1847–1849." *Humanities for the Environment: Integrating Knowledge, Forging New Constellations of Practice,* edited by Joni Adamson and Michael Davis, 181–194. Routledge, 2017.

Devine Guzmán, Tracy. *Native and National in Brazil: Indigeneity after Independence.* University of North Carolina Press, 2013.

Ferreira, Alexander Rodrigues. "Memória: A louça que fazem as índias do estado." In *Viagem ao Brasil de Alexandre Rodrigues Ferreira: Coleção Etnográfica,* edited by José Paulo Monteiro Soares and Cristina Ferrão, 3:75. Kapa, 2005.

Fonseca-Kruel, Viviane, Aloisio Cabalzar, Claudia Leonor López-Garcés, Márlia Coelho-Ferreira, Pieter-Jean van der Veld, Luciana Martins, William Milliken, and Mark Nesbitt. "Biocultural Collections and Participatory Methods: Old, Current, and Future Knowledge." *Methods and Techniques in Ethnobiology and Ethnoecology,* edited by Ulisses Paulino Albuquerque, Reinaldo Farias Paiva de Lucena, Luiz Vital Fernandes Cruz da Cunha, and Rômulo Romeu Nóbrega Alves, 215–228. 2nd edition. Humana/Springer, 2019.

Gan, Elaine, and Anna Tsing. "How Things Hold: A Diagram of Coordination in a Satoyama Forest." *Social Analysis* 62, no. 4 (Winter 2018): 102–145.

Gan, Elaine, Anna Tsing, Heather Swanson, and Nils Bubandt. "Introduction: Haunted Landscapes of the Anthropocene." In *Arts of Living in a Damaged Planet,* edited by Anna Tsing, Heather Swanson, Elaine Gan, and Nils Bubandt, 1–14. University of Minnesota Press, 2017.

Garfield, Seth. *In Search of the Amazon: Brazil, the United States, and the Nature of a Region.* Duke University Press, 2013.

Gonçalves, João. "Belém Declaration: Meat Industry Is Still off the Hook." *Mighty Earth,* 11 August 2023. https://mightyearth.org/article/belem-declaration-meat-industry-is-still-off-the-hook/.

Heckenberger, Michael J. "Mapping Indigenous Histories: Collaboration, Cultural Heritage, and Conservation in the Amazon." *Collaborative Anthropologies* 2 (2009): 9–32.

Hepp, Mauricio. "A emergência e dispersão do caraipé na cerâmica arqueológica da Amazônia e cerrado brasileiro: Temporalidade, relações sociais, identidade, resistência e cultura material." PhD dissertation, Universidade Federal de Minas Gerais, 2021.

Hooker, William. "Botanical Objects Communicated to the Kew Museum, from the Amazon River, in 1851, by Richard Spruce, Esq." *Hooker's Journal of Botany and Kew Garden Miscellany* 5 (1853): 169–177.

Hooker, William. "Botany." In *A Manual of Scientific Enquiry: Prepared for the Use of Officers in Her Majesty's Navy; and Adapted for Travellers in General,* edited by John F. W. Herschel, 400–422. John Murray, 1849.

Hooker, William. "Mr. Spruce's Intended Voyage to the Amazon River." *Hooker's Journal of Botany and Kew Garden Miscellany* 1 (1849): 20–21.

Hooker, William. "Mr. Spruce's Voyage to Pará." *Hooker's Journal of Botany and Kew Garden Miscellany* 1 (1849): 344–347.

Ingold, Tim. *Being Alive: Essays on Movement, Knowledge, and Description.* Routledge, 2011.

Lins, Juliana. *Cerâmica tukano.* Instituto Socioambiental, Federação das Organizações Indígenas do Rio Negro, 2020.

Martins, Luciana. "Plant Artefacts Then and Now: Reconnecting Biocultural Collections in Amazonia." *Mobile Museums: Collections in Circulation,* edited by Felix Driver, Mark Nesbitt, and Caroline Cornish, 22–43. University College London Press, 2021.

Martins, Luciana, and Lindsay Sekulowicz. "The Herbarium as a Workshop." *Ethnobotanical Assembly* 8 (2021). https://tea-assembly.com/issues/8/the-herbarium-as-a-workshop.

Mosse, Richard. "What the Camera Cannot See." *Art21,* 30 November 2022. https://art21.org/watch/extended-play/richard-mosse-what-the-camera-cannot-see/.

O'Brien, Phil. "Raymond Williams." In *The Bloomsbury Handbook of Literary and Cultural Theory,* edited by Jeffrey R. Di Leo, 738–739. Bloomsbury, 2018.

Oliveira, João Pacheco de. "Prefácio: Perda e superação." *No coração do Brasil: A expedição de Edgard Roquette-Pinto à Serra do Norte,* edited by Rita de Cássia Melo Santos, 7–23. Museu Nacional do Brasil, Setor de Etnología e Etnografía, 2020.

Oliveira, Thiago da Costa. *Cerâmica baniwa.* Instituto Socioambiental, Federação das Organizações Indígenas do Rio Negro, 2020.

Phillips, Tom. "Brazilian President Lula Pledges 'New Amazon Dream' at Rainforest Summit." *The Guardian,* 8 August 2023. https://www.theguardian.com/environment/2023/aug/08/brazilian-president-lula-pledges-new-amazon-dream-at-rainforest-summit.

Pitt Rivers Museum. "Biocultural Collections, Art, and Eco Activism: A Roundtable Discussion." Video. Posted to YouTube 22 September 2022. https://www.youtube.com/watch?v=p-eMxVH3caQ.

Plotkin, Mark J., Brian M. Boon, and Malorye Allison. Introduction to *The Ethnobotany of Aublet's* Histoire des Plantes de Guiane Française, edited by Mark J. Plotkin, Brian M. Boon, and Malorye Allison, 1–2. Volume 35 of *Monographs in Systematic Botany from the Missouri Botanical Garden.* 1991.

Rogero, Tiago. "Indigenous Leaders Frustrated Despite Cloak's Return to Brazil after 300 years." *The Guardian,* 16 July 2024. https://www.theguardian.com/world/article/2024/jul/16/brazil-indigenous-cloak-returned.

Rose, Deborah Bird. "Country and the Gift." *Humanities for the Environment: Integrating Knowledge, Forging New Constellations of Practice,* edited by Joni Adamson and Michael Davis, 33–44. Routledge, 2017.

Royal Botanic Gardens, Kew. "Pottery Trees." *Bulletin of Miscellaneous Information* 1903, no. 1:25–27.

Safier, Neil. "'Every Day I Travel . . . Is a Page That I Turn': Reading and Observing in Eighteenth-Century Amazonia." *Huntington Library Quarterly* 70, no. 1 (March 2007): 103–128.

Salick, Jan, Katie Konchar, and Mark Nesbitt. "Biocultural Collections: Needs, Ethics and Goals." In *Curating Biocultural Collections: A Handbook,* edited by Jan Salick, Katie Konchar, and Mark Nesbitt, 1–13. Kew, 2014.

Santos-Fonseca, Dyana Joy dos, Márlia Coelho-Ferreira, and Viviane Stern da Fonseca-Kruel. "Useful Plants Referenced by the Naturalist Richard Spruce in the 19th Century in the State of Pará, Brazil." *Acta Botanica Brasilica* 33, no. 2 (April–June 2019): 221–231. https:\\doi:10.1590/0102-33062018abb0344.

Schaller, Jörg, Benjamin L. Turner, Anita Weissflog, Delicia Pino, Aleksandra W. Bienilka, and Bettina M. J. Engelbrecht. "Silicon in Tropical Forests." *Biogeochemistry* 140, no. 2 (September 2018): 164–174.

Spruce, Richard. "Botanical Excursion on the Amazon." *Hooker's Journal of Botany and Kew Garden Miscellany* 2 (1850): 65–76.

Spruce, Richard. "Journal of an Excursion from Santarem, on the Amazon River, to Obidos and Rio Trombetas." *Hooker's Journal of Botany and Kew Garden Miscellany* 2 (1850): 266–276.

Spruce, Richard. "Journey of a Voyage up the Amazon and Rio Negro." *Hooker's Journal of Botany and Kew Garden Miscellany* 6 (1854): 33–42.

Spruce, Richard. Letter to Daniel Hanbury, 4 June [July?] 1865. At Royal Pharmaceutical Society.

Spruce, Richard. *Notes of a Botanist in the Amazon and Andes: Being Records of Travel on the Amazon and Its Tributaries, the Trombetas, Rio Negro, Uaupés, Casiquiari, Pacimoni, Huallaga, and Pastasa; as also to the Cataract of the Orinoco, along the Eastern Side of the Andes of Peru and Ecuador, and the Shore of the Pacific, during years 1849–1864.* 2 volumes. Edited and condensed by Alfred Russel Wallace. Macmillan, 1908.

Tupinambá, Glicéria (Célia). "The Cloak Is the Territory of the Tupinambá." Video. Posted to YouTube by University of St. Andrews School of Philosophical, Anthropological and Film Studies 4 November 2021. https://www.youtube.com/watch?v=Cnoy88m00nc&t=1053s.

Tupinambá, Glicéria (Célia). "An Indigenous Woman Troubling the Museum's Colonialist Legacy: Conversation with Glicéria Tupinambá." Interview. *Museum International* 74, no. 3–4 (2022): 10–23.

Varshney, Vibha. "Belem Declaration: Amazon Countries Fail to Agree on Protection Goals." *Down to Earth,* 9 August 2023, https://www.downtoearth.org.in/news/forests/belem-declaration-amazon-countries-fail-to-agree-on-protection-goals-91095.

Wallace, Alfred Russel. *A Narrative of Travels on the Amazon and Rio Negro, with an Account of the Native Tribes, and Observations on the Climate, Geology, and Natural History of the Amazon Valley.* Ward, Locke, 2nd ed., 1889 [1853].

Weinholt-Ludvigsen, Cecilie. "The National Museum of Denmark to Donate Rare Feather Cape to Brazil." *Via Ritzau,* 27 June 2023. https://via.ritzau.dk/pressemeddelelse/13700505/the-national-museum-of-denmark-to-donate-rare-feather-cape-to-brazil?publisherId=13560791.

Williams, Raymond. *Resources of Hope: Culture, Democracy, Socialism.* Edited by Robin Gable. Verso, 1989.

5

Poetics of an Expanded Humanity

FLORENCIA GARRAMUÑO

TRANSLATED BY NICOLÁS CAMPISI

When you feel the sky is caving in on you, just give it a push and breathe.
Ailton Krenak, *Ideas to Postpone the End of the World*

The past several decades have been fruitful for Amerindian cultural practices and knowledge, which have taken center stage in contemporary Latin America. These practices often appear as a vector of contesting forces and resistance knots that transform our understanding of Latin American culture. In *Planetary Longings,* Mary Louise Pratt reviews the conditions shaping this transformed "indigeneity" and the new role of Indigenous peoples during the end of the twentieth century and the beginning of the twenty-first, situating the neoliberal restructuring of the 1960s and 1970s as one of its most important catalysts. As Pratt points out,

> The biggest catalyst for global Indigenous activism was the new phase of capitalist expansion facilitated by free-trade agreements that favored multinational corporations and intensified extractive industry, especially in the Global South. The neoliberal restructuring of the global economy changed the game for Indigenous, tribal, and land-based peoples everywhere by placing their land bases in new jeopardy and weakening the ability of states to defend them against encroachment. (110)

In artistic practices, this force manifests in a commitment to a radical discussion of forms of capitalist and extractivist exploitation. Amerindian aesthetic practices question the vision and conception of "natural resources" that organized particularly violent forms of development in Latin America from the colonial era onward. These contemporary texts, installations, and videos feature animals, plants, mountains, and bodies of water. These beings are presented not as environment, medium, or landscape but in continuity with agencies and genealogies that overflow the limits of the contemporary Western imagination. What used to be called "nature" or "landscape" is absent from these new fictions and arts of the Anthropocene, as some scholars have named practices that engage with the discussion and elaboration of ways of feeling and perceiving the transformations of our contemporary worlds, characterized by species extinction and climate change (Davis and Turpin; Gan et al.; Trexler).[1] In these aesthetic practices, the deployment of a continuity between the nonhuman and the human and the staging of multiple genealogies that entangle human and geological times challenge the limits, forms, and concepts with which critics have traditionally conceived Latin American art and culture.

It is well known that across Latin America, especially in the Southern Cone and Brazil, the dictatorships of the 1960s and 1970s were the setting for the implementation of a fierce neoliberalism that not only produced an increasingly inequitable distribution of wealth but was sustained by a radical deepening of colonial extractivism (Svampa). In those decades, new Indigenous movements made their demands heard in various Latin American countries and powerfully made their ways of life visible (Danowski and Viveiros de Castro 119–148). James Clifford, in *Returns: Becoming Indigenous in the Twenty-First Century,* studies the resurgence of Indigenous movements during recent decades throughout the planet. He notes the convergence of Indigenous resurgence and neoliberalism in the 1980s and 1990s, emphasizing that it could not be explained through aesthetic or historical markers like "late capitalism" or "postmodernity." Clifford contends, "The convergence cannot be rounded up with periodizing terms like 'late capitalism' or 'postmodernity.' Nor can we draw a simple link between political-economic structures and sociocultural expressions, claiming that one element (in this case, indigenous resurgence) is a result, or a production, of the other (neoliberal hegemony)" (16). Amerindian intellectual struggles and practices began much earlier, as Héctor Nahuelpan Moreno has shown (17). The Mapuche artist Seba Calfuqueo has observed that Mapuche art has always existed, but it was not seen as contemporary or made visible as such. In fact, some still consider the category "Mapuche contemporary art" to be a novelty, an art that only now emerged from the nothingness of an empty tradition (Calfuqueo).

Beyond this long persistence, it is evident that Amerindian cultural practices and the increased presence of Indigenous peoples as political subjects in various Latin American societies have acquired new and surprising visibility and relevance in current debates. Helen Gilbert has remarked,

> The view that indigeneity underpins contemporary debates of global importance has broad resonance across a number of scholarly disciplines and in the material, artistic, and ideological domains they map. There has been a marked increase over the last ten to twenty years in the visibility and reach of cultural and political networks invested in indigeneity, especially in Latin America. Also, questions about what particular rights that status might confer in relation to natural resources, cultural practices, governance, and sovereignty, to name just a few contentious areas, have spread well beyond the confines of nation-states tasked with the unfinished business of decolonization. (174)

I believe that this new relevance of Amerindian practices has to do, among other reasons, with how they articulate a powerful critique of forms of capitalist domination that the neoliberal reforms of recent decades have laid bare in all their violence. These practices allow, in different ways, a review of the history of capitalist exploitation that implanted in Latin America the project of extinction not only of Amerindian populations but also of their knowledge, practices, and cosmogonies. In texts, installations, videos, visual works, and poems, contemporary Amerindian art incites an evident displacement of the human figure in favor of mineral, vegetal, and animal voices and bodies. These strategies compel us to imagine less destructive relationships with other species and the environment, introducing into the contemporary imagination other possible futures. In this chapter I analyze the work of Seba Calfuqueo and the writings of Ailton Krenak and the way their imagination of natural elements in continuity with human agents displaces the dissociative vision of natural sciences and contests the extractive impulse that supported that vision. They both react to and contest that vision with texts and art installations, historical methods of displaying and conceiving nature, and postanthropocentric perspectives. Other contemporary artists engage critically and creatively with older genres and archives of the natural sciences. In contrast, Calfuqueo and Krenak, along with an increasing number of Indigenous artists and intellectuals, draw attention to Indigenous epistemologies and cosmogonies to imagine different and horizontal relations between humans and nonhuman agents.

The Forms of Water

In the context of the water crisis that has been worsening in Chile in recent years, the 2015 project to build a water highway that would take water from the southern lakes to the north of the country to remedy the drought there, in a measure proposed as "interregional solidarity" (*solidaridad interregional*), provoked a series of intense debates and various actions by activists (Bogliolo).[2] The water code passed during the Pinochet dictatorship in 1981 conceives water as a tradeable good in the market and supports the proposal. Some scientists have questioned the project, warning that it would irreversibly damage the ecosystems and natural life that depend on these flows. It is in this context that Calfuqueo had an exhibition titled *Ko ñi Weychan* (The Struggles of Water) on March 14–15, 2020, at the Galería Metropolitana in Santiago, Chile. The exhibition's curator, Cristian Vargas Paillahueque, offers a critique of the water crisis in an article published in November 2020.

> The water crisis in Chile has worsened over the last decades, and its impact on the Indigenous people has increased. Avocado and vineyard plantations in the central zone and monoculture forestry plantations in the south have caused real damage to biodiversity and life in the communities. We must also add the predatory eagerness of hydroelectric or mining projects that prioritize commercial interests over the dignity of the people. From the conception of *itrofill mongen* (distinct lives), this is increased, given that water scarcity influences species preservation and implicates the drought of spaces where, for example, medicinal herbs are obtained. Episodes involving the monumental use of water for transnational companies have marked the imaginary of the struggles against violent forms of extractivism, as was the installation of the Ralco dam that made visible the struggles of the Quintreman sisters or, even more recently, the death of Macarena Valdés for defending the Trangil space, and whose death is still being investigated as a corporate hitman assassination. The hegemonic press has been feasting on the map of territorial conflicts, ignoring the role of the companies and the political class in this depredation that puts market rules above those affecting us all.

> La crisis del agua en Chile se ha acrecentado durante las últimas décadas y el impacto ha sido cada vez mayor sobre los pueblos. Monocultivos de plantaciones de paltos y viñedos en la zona central y monocultivos forestales, en el sur, han causado verdaderos estragos en la biodiversidad y en las poblaciones. A eso debemos sumar el afán depredador de hidroeléctricas o proyectos mineros que ponen intereses comerciales por sobre la dignidad de la gente. Desde la concepción de itrofill mongen (distintas vidas) esto se acrecienta, dado que la escasez de agua influye en la preservación de especies e implica la sequía de espacios donde, por ejemplo, se obtienen yerbas medicinales. Episodios que im-

> plican el uso monumental de aprovechamiento de aguas para empresas transnacionales han marcado el imaginario de las luchas desde las formas más violentas, como lo fue la instalación de la represa de Ralco que visibilizó las luchas de las hermanas Quintreman o, más recientemente aún, la muerte de Macarena Valdés por defender el espacio de Trangil, y cuya muerte sigue siendo investigada por asesinato del sicariato empresarial. La prensa hegemónica ha festinado sobre el mapa de los conflictos territoriales ignorando el rol de las empresas y de la clase política en esta depredación que pone las reglas del mercado por sobre las de todos. (Vargas Paillahueque)

Calfuqueo's exhibition consisted of two interrelated works: *Kowklen* (In a Liquid State) and *Ko ta mapungey ka* (Water Is Also Territory), using two of the most paradigmatic languages of contemporary art—installation and performance—to contextualize the problem of the lack of water. Each work displays the universe of Mapuche memory and ontology for which water is sacred and considered a living being. Vargas Paillahueque explains,

> The exhibition *Ko ñi weychan* (The Water Struggles) by Sebastián Calfuqueo mounts a reflection that makes the water issue visible in the Chilean neoliberal context. The strategy that frames the exhibition in Galería Metropolitana (Santiago) consists of vindicating Wallmapu [the historical territory inhabited by the Mapuche Indigenous people] as a space of enunciation to raise a critical voice, bringing to the fore a necessary decolonizing approach that such a task demands when putting in dialogue the resistance of the rivers with a model that is in crisis.

> La exposición *Ko ñi weychan* (Las luchas del agua) de Sebastián Calfuqueo instala una reflexión que permite visibilizar la problemática de las aguas en el contexto neoliberal chileno. La estrategia que enmarca la muestra en Galería Metropolitana (Santiago) consiste en reivindicar el Wallmapu como espacio de enunciación para plantear dicha crítica, advirtiendo con ello un necesario enfoque descolonizador que exige dicha tarea cuando se pone en diálogo la resistencia de las aguas frente a un modelo en crisis.

A few months later, in 2021, a new constitutional convention was discussing, among other issues, the right to water for all peoples based on the Indigenous cosmovision and knowledge whose recognition that same constituent body aimed to extend. During that time, Calfuqueo exhibited, in the same gallery, the show entitled *Espejo de agua* (Water Mirror), made up of two works: *Mercado de aguas* (Water Market) and *Palabras al agua* (Words to Water). *Mercado de aguas* displays a series of pieces of blue ceramic made from a plastic container of 20 liters of water each. In the installation, each bright-blue glazed ceramic drum contains a phrase from the 1981 Water Code of Chile inherited from the dictatorship: "Derecho de aprovechamiento" (Right of Use), "Decreto" (Decree), "Extracción" (Extraction). The water packaged for commercialization in plastic containers and the neoliberal measures to which these words point contrast with others such as "Cascadas" (Waterfalls), "Lluvias" (Rain), and "Pantanos" (Swamps), about the ecosystems that nest around the free waters of the Mapuche Indigenous community. The Chilean Water Code, the forms of water distribution in plastic containers, and neoliberalism are contrasted in the installation with the language of ceramics, an ancestral Mapuche practice, and with the color blue, which is sacred in Mapuche culture. Drums placed on tables read, "Código es saqueo" (The code is a plunder), "Uso público como propiedad" (Public use as property), "Lagos libres" (Free lakes), and "Bienes para todes" (Goods for all).

The second installation, *Palabras al agua,* consists of blue canvas that descends from the wall and vibrates through the space like a river between stones and ceramic pieces with words written on them that allude to the free-flowing waters. From some stones comes the sound of poems about water. With elements that symbolize her Indigenous ancestry and her conversation with contemporary art, Calfuqueo highlights the intersection and overlapping of ancestral and contemporary issues, displaying the power of Indigenous knowledge and culture to question the extractive practices that are at the roots of the Anthropocene. In a text about the installation, María José Barros observes,

> According to the intellectual and ecofeminist activist Vandana Shiva in her book *Water Wars,* the market paradigm conceives water as an inert and feminized body that can be subject to colonization, exploitation, and commercialization. In contrast, the ecological paradigm understands water as a gift granted by nature and a communal good that must be cared for and distributed equitably. . . . In the case of Calfuqueo, the disagreement between both ways of thinking, feeling, and relating to this vital liquid in dispute today is represented from a perspective rooted in Mapuche epistemology and, at the same time, from a nonbinary and performative identity position related to the fluid materiality of water. Hence, the possibility of thinking of this installation as a manifestation of recent Indigenous artistic activism, which seeks to promote the decolonization of knowledge, bodies, and nature, but also of Chilean art institutions, still marked by classist and racist practices.

> De acuerdo con la intelectual y activista ecofeminista Vandana Shiva en su libro *Las guerras del agua*, el paradigma de mercado concibe el agua como un cuerpo inerte y feminizado que puede ser objeto de colonización, explotación y comercialización, mientras que el paradigma ecológico entiende el agua como un don gratuito otorgado por la naturaleza y como un bien comunitario que debe ser cuidado y distribuido equitativamente. . . . En el caso de Calfuqueo, el desencuentro entre ambas formas de pensar, sentir y relacionarse con este líquido vital hoy en disputa es representado desde una perspectiva enraizada en la epistemología mapuche y, al mismo tiempo, desde un posicionamiento identitario no binario y performativo afin a la materialidad fluida de las aguas. De ahí la posibilidad de pensar esta instalación como una manifestación del activismo artístico indígena reciente, que busca promover la descolonización de los saberes, los cuerpos y la naturaleza misma, pero también de la institucionalidad del arte chileno, aún marcada por prácticas clasistas y racistas. (Barros)

The different installations embody an insistence on the problem of water and a way of inserting themselves time and again in contexts of ecological struggles in which the languages of contemporary art intertwine with those of ancestral cultures. In this sense, Calfuqueo replaces water as a consumer good with water as a living and fluid being. This Mapuche conception of water enters the institutional space of art in these installations, displacing the idea of water—or rivers and lakes—as an object of representation, making the museum and the gallery thus become, more than exhibition spaces, sites for reflection. Indigenous epistemologies inflect new ways of thinking about diverse life forms and the different ways in which we can respect them.

Other Epistemologies

Ailton Krenak's *Ideas to Postpone the End of the World* also mobilizes ancestral knowledge of Indigenous cultures to intervene in central issues of the contemporary debate on nature and the future. In three intertwined lectures, Krenak insists on developing a continuity between nature and humanity that disarms anthropocentric hierarchies with the perspectivist imagination characteristic of Amerindian peoples.[3] I want to dwell on the materiality of this text, which, together with the other texts published so far by this intellectual of the Krenak ethnic group, turn what were originally conversations or lectures into a very particular textuality, interwoven with dialogues and debates, and whose dialogic disposition incorporates ideas from diverse places to weave a mesh that integrates contrasting visions. I am also interested in focusing on the literary condition of this text beyond its argumentative dimension and its ideas because I believe that in this dialogic coexistence shines an aesthetic form that does not differentiate argument and narrative. Instead, the text approaches the argument from its narrative configuration, or, in other cases, it is the narrative that introduces the development of the argument. The text is interwoven with debates and conversations with some of the most influential intellectuals of contemporary thought, such as Boaventura de Sousa Santos, Davi Kopenawa, and Eduardo Viveiros de Castro. In these discussions, the questioning of the idea of Western humanity responsible for the Anthropocene is concurrent with the construction of an "us" that includes mountains, rivers, and animals as part of the same continuum. For Krenak,

> The Doce River, which we, the Krenak Nation, call Watu—our grandfather—is a person, not a resource, as the economists like to call him. He is not something you can own or appropriate; he is part of our construction as a collective society that dwells in a specific place into which we have been gradually corralled by the government, forcing us to live and breed in bubbles subject to increasingly crippling external pressure. (43–44)

That same river was contaminated with toxic materials spilled from mining exploitation due to a dam rupture in Mariana, Minas Gerais. Faced with this context of destruction, Krenak's ideas reflect on our contemporary world with images of an environment that appears as a living being with agency and that is presented in continuity with the human figure. As in the demonym Krenak—constituted, as Ailton reminds us, by the prefix *kre* (head) and *nak* (earth)—the Krenak would be the head of the Earth in a continuity where no verticality points to hierarchies or domination. We should not separate humans from nature, Krenak warns, "as if we were not nature ourselves" (65).

In his afterword for the French and Portuguese editions of *Ideas to Postpone the End of the World*, Viveiros de Castro proposes Krenak's book as proof of the failure of a specific idea of humanity, which "by having placed the metaphysical devaluation of the world as its very condition of possibility, transformed the bearers of this idea into agents of the physical destruction of this same world" [ao ter posto a desvalorização metafísica do mundo como sua própria condição de possibilidade, transformou os portadores dessa ideia em agentes da destruição física deste mesmo mundo] ("Posfácio" 79). In Krenak's view, "This humanity refuses to recognize that the river, now in a coma, is also our grandfather; that the mountains mined in Africa or South America and transformed into merchandise elsewhere are also the grandfather, grandmother, mother, brother of some other constellation of human beings that want to go on sharing the communal home we call Earth" (*Ideas* 49). Against this conception of humankind, Krenak proposes the idea of an expanded humanity, sustained in a continuity between nature and humans that results in the articulation of forms of resistance and critique of contemporary neo-extractivism.

In Krenak's later book, *Futuro ancestral* (Ancestral Future), the importance of this conception of rivers as free and living waters reinforces how bodies of water are beings with agency when considered from an Amerindian perspective. For Krenak, "The rivers, those beings that have always inhabited different worlds, are the ones that suggest to me that if there is a future to imagine, it is ancestral, because it is already present" (*Ancestral Future* 1). In this extension of the notion of living beings and agency to rivers and natural environments, Krenak's thought unfolds the epistemology of his people, illuminating the contemporary moment and its crises with a new light.

Supernature and Perspectivism

In Krenak's texts and Seba Calfuqueo's installations, these poetics of an expanded humanity can be productively read through the category of *sobrenatureza* (supernature) proposed by Viveiros de Castro. In Marco Antonio Valentim's interpretation of the concept, "Supernature seems to consist in the perspective itself, an element that, destabilizing the division between nature and culture, establishes the dynamics of alteration and 'symmetrical reciprocity' between them" [A sobrenatureza parece consistir na própria perspectividade, elemento que, desestabilizando a divisão entre natureza e cultura, instaura a dinâmica de alteração e "reciprocidade assimétrica" entre elas].[4] This also explains the figurations without hierarchies that appear in the texts and installations studied here. Valentim articulates a counterontology by contrasting the concept of being in Western and Amerindian philosophies. I am interested in underlining the production of concepts and not only of new sensibilities that these aesthetic and intellectual practices generate with their knowledge. These artists and practitioners explore sensibilities and affections that expand the notion of living beings. By reflecting on our Anthropocene present, Krenak's writings and Calfuqueo's installations on water coincide in identifying a series of Amerindian knowledge systems as ways of opening time horizons that, silenced in the past, gain an essential power of reactivation.

These practices inspire us to review the history of capitalist exploitation, a project of European colonization that continued with the formation of modern nation-states (Cárcamo-Huechante). In the context of neoliberalism and neo-extractivism of the first decades of the twenty-first century in Latin America, the reflection and resistance these practices exhibit compel us to imagine other ways of relating to the rest of the living and nonliving beings of the planet. These poetics of an expanded humanity emerge as a way of thinking about the present and warning about the risks and the difficulties that the narrow conception of humanity characteristic of Western thought has brought to the planet. Pedro de Niemeyer Cesarino has remarked about Davi Kopenawa and Bruce Albert's *The Falling Sky: Words of a Yanomami Shaman*, another generous and powerful writing by a contemporary Indigenous thinker,

> *The Falling Sky* is not only . . . the most important text produced recently by an Amerindian, as if its relevance could be delimitated only by an ethnographical scope; it is also one of the most powerful contemporary narratives about the radical reconfiguration of the earth-system by human agency or, in other words, the Anthropocene. . . . Its originality, however, does not rest exactly on the diagnosis of chaos but rather on the perseverant offering of another possible world and condition of humanity. (294)

The offering of another possible world, manifested in the words of Kopenawa by a desire "to warn the white people before they wind up tearing the sky's roots out of the ground" is configured in the works of Calfuqueo and Krenak from these poetics of an expanded humanity (314). These Indigenous thinkers provide a powerful device for imagining other possible futures.

Notes

1 I use "Anthropocene" as a term with an open meaning that has been the subject of several controversies. Proposed by Paul Crutzen in 2002 to define the geological era in which humans have become the major force determining the continued possibility of life on the planet (Gan et al. 24), the concept has brought together a series of debates about its meaning and the appropriateness of naming the epoch by this category or others. I am interested in taking the Anthropocene as a figure of the epistemological struggles that the very notion of a new geological age raises. For Elaine Gan, Anna Tsing, Heather Swanson, and Nils Bubandt, "Our use of the term 'Anthropocene' does not imagine a homogeneous human race. We unite in dialogue with those who remind readers of unequal relations among humans, industrial ecologies, and human insignificance in the web of life by writing instead Capitalocene, Plantacionocene, or Chutululucene" (3).

2 Unless otherwise indicated, all translations from Portuguese and Spanish are the translator's own.

3 Amerindian perspectivism is defined by Viveiros de Castro in the following way: "Typically, humans, under normal conditions, see humans as humans, animals as animals and spirits (if they see them) as spirits; while animals (predators) and spirits see humans as animals (of prey), while animals (of prey) see humans as spirits or as animals (predators). In return, the animals and spirits see themselves as humans: they apprehend themselves as (or become) anthropomorphic when they are in their own homes or villages and experience their own habits and characteristics under the species of culture—they see their food as human food . . . , their bodily attributes . . . as adornments or cultural instruments, their social system as organized in the same way as human institutions. . . . This 'seeing as' refers literally to percepts and not analogically to concepts, even if, in some cases, the emphasis is more on the categorical than the sensory aspect of the phenomenon; in any case, the shamans, masters of cosmic schematism . . . , dedicated to communicating and managing these crossed perspectives, are always there to make concepts sensible or intuitions intelligible" [Tipicamente, os humanos, em condições normais, vêem os humanos como humanos, os animais como animais e os espíritos (se os vêem) como espíritos; já os animais (predadores) e os espíritos vêem os humanos como animais (de presa), ao passo que os animais (de presa) vêem os humanos como espíritos ou como animais (predadores). Em troca, os animais e espíritos se vêem como humanos: apreendem-se como (ou se tornam) antropomorfos quando estão em suas próprias casas ou aldeias, e experimentam seus próprios hábitos e características sob a espécie da cultura vêem seu alimento como alimento humano . . . , seus atributos corporais . . . como adornos ou instrumentos culturais, seu sistema social como organizado do mesmo modo que as instituições humanas. . . . Esse "ver como" se refere literalmente a perceptos, e não analogicamente a conceitos, ainda que, em alguns casos, a ênfase seja mais no aspecto categorial que sensorial do fenômeno; de todo modo, os xamãs, mestres do esquematismo cósmico . . . , dedicados a comunicar e administrar essas perspectivas cruzadas, estão sempre aí para tornar sensíveis os conceitos ou tornar inteligíveis as intuições] ("Os pronomes").

4 On the concept of *sobrenatureza,* see also Viveiros de Castro's *A inconstância da alma selvagem;* an abridged version of the book was translated as *The Inconstancy of the Indian Soul: The Encounter of Catholics and Cannibals in 16th-Century Brazil.* In the book's Brazilian edition, Viveiros de Castro remarks that "the typical supernatural situation in the Amerindian world is the encounter in the forest between a human being, always alone, and a being who, first seen as an animal or a person, reveals itself as a spirit or a dead person, and speaks to the human. . . . These encounters are lethal for the interlocutor who, subjugated by the nonhuman subjectivity, crosses over to its side, becoming a being of the same species as the speaker" [A situação sobrenatural típica no mundo ameríndio é o encontro, na floresta, entre um humano—sempre sozinho—e um ser que, visto primeiramente como um mero animal o uma pessoa, revela-se como um espírito ou um morto, e fala com o homem. . . . Esses encontros costumam ser letais para o interlocutor, que, subjugado pela subjetividade não-humana, passa para o lado dela, transformando-se em um ser da mesma espécie que o locutor] (397).

Works Cited

Barros, María José. "*Espejo de agua* (2021) de Sebastián Calfuqueo: Aguas libres, identidades fluidas." *Endémico,* May 18, 2022. https://endemico.org/espejo-de-agua-2021-de-sebastian-calfuqueo-aguas-libres-identidades-fluidas/.

Bogliolo, Félix. "Trasvasije de agua desde sur al norte de Chile: Una acción de solidaridad interregional." *Ciper,* August 12, 2021. https://www.ciperchile.cl/2021/08/13/trasvasije-de-agua-desde-sur-al-norte-de-chile-una-accion-de-solidaridad-interregional/.

Calfuqueo, Seba. "Entrevista/Interview." November 26, 2020. 34th Bienal de São Paulo. http://34.bienal.org.br/en/artistas/8285.

Cárcamo-Huechante, Luis. "A Trope of Colonial Obliteration? A Critical Note on 'Colonial Latin America' and Related Conversations." *Colonial Latin American Review* 32, no. 2 (2023): 243–248.

Clifford, James. *Returns: Becoming Indigenous in the Twenty-First Century.* Harvard University Press, 2013.

Danowski, Déborah, and Eduardo Viveiros de Castro. *¿Hay mundo por venir? Ensayo sobre los miedos y los fines.* Caja Negra, 2017.

Davis, Heather, and Etienne Turpin. *Arts in the Anthropocene: Encounters among Aesthetics, Politics, Environments, and Epistemologies.* Open Humanities, 2015.

Gan, Elaine, Anna Tsing, Heather Swanson, and Nils Bubandt. "Introduction: Haunted Landscapes of the Anthropocene." In *Arts of Living on a Damaged Planet: Ghosts of the Anthropocene,* edited by Gan, Tsing, Swanson, and Bubandt, 1–14. University of Minnesota Press, 2017.

Gilbert, Helen. "Introduction: Indigeneity and Performance." *Interventions* 15, no. 2 (2013): 173–180.

Kopenawa, Davi, and Bruce Albert. *The Falling Sky: Words of a Yanomami Shaman.* Translated by Nicholas Elliott and Allison Dundy. Harvard University Press, 2013.

Krenak, Ailton. *Ancestral Future.* Translated by Alex Brostoff and Jamille Pinheiro Dias. Polity, 2024.

Krenak, Ailton. *Ideas to Postpone the End of the World.* Translated by Anthony Doyle. Anansi International, 2020.

Nahuelpan Moreno, Héctor, Herson Huinca Puitrin, Pablo Mariman Quemenado, Luis Cárcamo-Huechante, Maribel Mora Curriao, José Quidel Lincoleo, Enrique Antileo Baeza, et al. *Ta iñ fijke xipa rakizuameluwün. Historia, colonialismo y resistencia desde el país Mapuche.* Ediciones Comunidad de Historia Mapuche, 2012.

Niemeyer Cesarino, Pedro de. "Ontological Conflicts and Shamanistic Speculations in Davi Kopenawa's *The Falling Sky.*" *Hau: Journal of Ethnographic Theory* 4, no. 2 (2014): 289–295.

Svampa, Maristella. *Neo-Extractivism in Latin America: Socio-Environmental Conflicts, the Territorial Turn, and New Political Narratives.* Cambridge University Press, 2019.

Page, Joanna. *Decolonial Ecologies: The Reinvention of Natural History in Latin American Art.* Open Book, 2023.

Pratt, Mary Louise. *Planetary Longings.* Duke University Press, 2022.

Trexler, Adam. *Anthropocene Fictions: The Novel in a Time of Climate Change.* University of Virginia Press, 2015.

Valentim, Marco Antonio. "Sobrenatureza: Antropomorfia ou monstruosidade?." Paper presented at XI Congreso Argentino de Antropología Social, Rosario, 2014. https://cdsa.aacademica.org/000-081/1318.

Vargas Paillahueque, Cristian. "*Ko konümpakey tañi weychan* (el agua rememora sus luchas). Sobre *Ko ñi weychan,* de Sebastián Calfuqueo." *Artishock: Revista de Arte Contemporáneo,* November 28, 2020. https://artishockrevista.com/2020/11/28/ko-ni-weychan-sebastian-calfuqueo/.

Viveiros de Castro, Eduardo. *A inconstância da alma selvagem (e outros ensaios de antropologia).* Cosacnaify, 2002.

Viveiros de Castro, Eduardo. "Os pronomes cosmológicos e o perspectivismo ameríndio." *Mana* 2, no. 2 (1996): 115–144.

Viveiros de Castro, Eduardo. "Posfácio: Perguntas inquietantes." In *Ideias para adiar o fim do mundo,* by Ailton Krenak, 73-84. Companhia das Letras, 2020.

6

Telling Natural Histories in Crispín Amador Ramírez's *El infierno del paraíso* and Kali Fajardo-Anstine's *Woman of Light*

Matylda Figlerowicz

Museums and novels partake in modern ways of imagining the world. They tell stories meant to represent a larger whole, an imagined community. Discourses built upon museums and novels have played a significant role in the perception of Indigenous populations within the world that has deemed itself modern. In this essay I examine the intertwining of natural history and storytelling in two novels: Crispín Amador Ramírez's *El infierno del paraíso* (2005) and Kali Fajardo-Anstine's *Woman of Light* (2022). I analyze, thus, twenty-first-century Indigenous narratives from different contexts—a novel by a Nahua writer that came out in a bilingual edition in Nahuatl and Spanish and a novel written in English by an Indigenous Chicana author. As I study the two novels, I ask how telling stories interweaves with developing different forms of care for the nonhuman and human world at the threshold of the third millennium.

In both novels we find an explicit treatment of these subjects; the characters are confronted with and involved in structures of exploitation of the human and nonhuman world. The novels can be read as spaces for denouncing extractivism as well as for preserving the memory or mourning the loss of species and their interrelationships. But in less overt ways the novels also impel us to ask what practices of storytelling can destabilize the scientific and museographic discourses of natural history, as the writers experiment with forms of undermining human-centered images of the natural world and its transformations. I seek to examine how these twenty-first-century authors, in different contexts and languages, try out strategies to test the extent to which the novel can become a text penetrated by nonhuman agents.

The novels seem to ask about the possibilities of today's cultural production: whether it can give rise to aesthetic and epistemological forms that make more space for the nonhuman or even put the natural world in control of a narrative. The texts provide different answers to these questions. We will often see them fail, or at least respond negatively; many times it seems like the engagement with any organic and inorganic matter is so saturated with human perspective that nature can only take shape as an object of people's observation and interaction. In these moments of the novels, the practice of taking care through telling stories once again turns into taking control. These instances are stark reminders of how every discourse of natural history tells also (and sometimes primarily) the story of its narrators. Yet other times, spaces open up where giving up control over the narrative becomes conceivable, as novelistic storytelling strives to put in motion more radical modes of care for the human and nonhuman world.

*

Natural history museums arise from institutions that are not at first quite as clearly delimited. If we take as an example the Natural History Museum in London, we see how its story is entwined with that of the British Museum, which to this day is described on its webpage through a desire for an exhaustive vision and reach:

> The British Museum was founded in 1753 and opened its doors in 1759. It was the first national museum to cover all fields of human knowledge, open to visitors from across the world. . . . The Museum is driven by an insatiable curiosity for the world, a deep belief in objects as reliable witnesses and documents of human history, sound research, as well as the desire to expand and share knowledge.

Hans Sloan's collection, which was "the catalyst that brought the British Museum into being" (Thackray and Press 19), included "plants, corals, minerals, earth, shells, animals, insects etc.," as well as "antiquities of all sorts, coins and medals, prints and drawings, books and manuscripts, and what we would now class as ethnographic collections" (14). After decades of expanding the vast and multifaceted collection, the natural history branch of the museum separated in 1881.

In the British Museum we find a gallery room celebrating the museum's own history and the collectors who contributed to it, called Collecting the World. Natural history discourses are rooted in this broader impulse of gathering, describing, and displaying elements of the human and nonhuman world that would allow the collectors and the visitors to have a feeling of having touched or even grasped the world. The history behind such collections has been under increasing scrutiny over the past several decades as a history of theft, destruction, and disrespect. Writing about Indigenous peoples' relations with museography, Philip Deloria draws a genealogy starting with the Renaissance cabinets of curiosities, which served as material proof of the European elite's colonial expansion and imperial power (107). Deloria analyzes how these expanding collections soon made Indigenous peoples themselves into objects of study, enclosed in reductive scientific categories—natural history, ethnology, anthropology, archaeology, and craniology—where they served as a backdrop against which Western culture understood itself as more advanced, modern, civilized, and ultimately, more human (108).

Many museums and research practices are still based on such conceptualizations of Indigenous peoples and their cultures, and to this day, in visiting some natural history museums we can find exhibitions dedicated to Indigenous peoples' past or present.[1] In these cases, the ways in which museums collect and contextualize Indigenous objects continue to correspond with the reductive disciplinary pigeonholing that assumes that Indigenous cultures are relevant only to a narrow set of questions and historical moments.[2] Toward the end of the twentieth century, important changes took place in legislation and in museum practice.[3] As Deloria analyzes the foundation of the National Museum of the American Indian in 1989, he describes its nonlinear and multivocal structure as an example of how Indigenous curatorship can create "the possibilities of a postcolonial and postmodern practice" (111).

I am interested in similar shifts that we can observe in Indigenous novelistic production. More specifically, I ask how Amador Ramírez and Fajardo-Anstine use the novel to engage with discourses that, throughout centuries, have participated in shaping both the Western perception of Indigenous peoples and the ways in which human and nonhuman stories intertwine. The novel as a genre is understood as a literary form deeply tied to modernity and its global reach. Mariano Siskind notes,

> During the second half of the nineteenth century, when bourgeois reason (through its economic, political, and cultural institutions) was believed to occupy every region of the planet, the novel produced privileged and efficient narratives of global formation of a modern world. Because the novel was the hegemonic form of narrative imagination in the nineteenth century, and because of the aesthetic and political force of its social totalities, most novels dealing with distant places generated powerful images of the globalization of modern culture. (26)

Indigenous people have been treated as themes and objects of novelistic representation rather than their authors and audience. More than that, their literary production would commonly be defined almost in opposition to the novelistic genre. Gloria Chacón writes that the widespread belief was that "indigenous people have not and do not produce novels, and that it is precisely this genre that nonindigenous intellectuals used to narrate the disintegration of indigenous cultures" (127). Contemporary Mayan and Zapotec novels, Chacón argues, become a "powerful tool" in opposing absorption and projecting a future of autonomy "as a labor that lies ahead in order for indigenous nations to realize self-sufficiency" (151). Nevertheless, as this image of possible liberation appeals to the concept of nation, it relies on another modern ideological formation.

While the novel is a genre with strong connections to modernity, globalization, and national formation, its position within today's cultural panorama and sociopolitical imagination is different than in the nineteenth and early twentieth centuries. The increasing expansion of global capitalism and the exploitation of natural resources is accompanied by changes in forms of communication and cultural production that hold primacy over the most common images of the world. Similarly, natural history discourse does not play the same role as it had in the modern colonial expansion, yet it is important to a world living through climate change. It is relevant for its desire to safekeep and protect nature, but it also finds itself under renewed scrutiny for its participation in imperialist projects, commonly resulting in further abuse rather than preservation of the nonhuman world.

The intercrossing of the novelistic genre and natural history is a site where we can reflect on the forms of thought that shaped modern and colonial projects. Tracing the genealogy of natural history shows that the domination over the natural world depends on the same power dynamics as the global imperial structures. While to this day, such a sense of ownership over nature keeps haunting natural history museums, the discourse of natural history also expresses the desire for preservation and care. It is rooted in the fascination with the organic and inorganic world through the attention paid to its particular representatives. The novels that I read in this essay share this fascination and try to imagine narrative forms that would respond to a more porous perception of the boundaries between the human and nonhuman.

*

A museum of natural history might exhibit plants or animals—already lifeless—that reflect current ecosystems but also speak to how they have been changing and preserve knowledge of what has been lost. There is something of a similar desire within Amador Ramírez's *El infierno del paraíso*, where different forms of the organic and inorganic world come to the fore as they are threatened by capitalist exploitation of natural resources; in Arturo Arias's words, "The exterior details are not decorative. They are in the spotlight. Sometimes they are the substitute of action itself, which often is relegated to the background" [Los detalles externos no son decorativos. Son el primer plano. A veces, son el sustituto de la acción misma, la cual queda relegada con frecuencia al fondo escénico] (164).[4] As the novel progresses, however, we observe how the natural histories that are tied to the protagonist's story become increasingly untellable. The severed relationship with land and its nonhuman inhabitants makes it impossible to narrativize natural history, understood as the story of the organic and inorganic matter that makes up the world and as the story of the narrator constructing this account. At the same time, the novel attempts to leave space for what cannot be told anymore, gesturing toward the possibilities of storytelling that are lacking because of the hostility marking the relationship to the land.

The text deautomatizes the ways in which we tell our stories by making literary conventions unobvious. *El infierno del paraíso* is considered to be the first novel in Nahuatl, and it experiments with the genre in terms of its position within the Nahua literary tradition and in terms of the novelistic form. The novelistic tradition has been perceived as at odds with Indigenous cultures, and by writing a novel in Nahuatl, Amador Ramírez poses the question of how the genre can be transformed in contact with other Indigenous literary practices. This first novel in Nahuatl is actually two novels within one; on each left-hand page we get the version in Spanish, and on the right-hand one, the version in Nahuatl. In what ways are these complementary accounts, and in what ways are they in conflict? We also encounter two voices in the narration. While the novel starts off with a third-person narrator, it switches back and forth into a first-person narration of the protagonist, Mundo. It does not seem to follow a clear pattern and the shifts are not announced in any way, but in the beginning of the last chapter the novel settles into Mundo's first-person narration.

The opening pages of *El infierno del paraíso* speak of climate change and unequal land distribution. Mundo worries how to get by without owning much property, especially as his land is yielding fewer crops due to the changing ecosystem. The novel goes on to tell us how he and his friend Nabor take up seasonal jobs on different plantations. The novel is structured in three chapters, each of which is centered around a seasonal job Mundo and Nabor go on. The first two chapters cover the two jobs that begin their life as migrant workers; the last chapter jumps in time and summarizes the many jobs they take up over the years, to then focus on the last job Mundo and Nabor depart on together.

The way in which the work on the plantations is depicted in the novel draws a parallel between the mistreatment of land and of people. Monocultures go together with the concentration of land in the hands of individual owners and the abuse of workers. The destruction of ecosystems and local communities is part of the same process of economic exploitation. Mundo and Nabor's temporary jobs as migrant workers influence their daily life in their community. While they reach more economic stability thanks to the extra income, exceeding what they would be able to earn within their village, the relationships with the nonhuman and human world that define their lives and values are shaken.

Each time Mundo and Nabor depart for a new job, they are not sure of the conditions they will encounter. The accommodations and food they receive are precarious. Some workers die on the plantations either because of the dangerous work conditions or because they are killed by overseers. From the very first description of the work at a plantation, Mundo is depicted as strongly affected by the hostile conditions to which he is exposed:

> soon the machete started to hurt him, and he urinated on his hands, thus avoiding the forming of blisters, around midday the temperature was unbearable, the hot air almost burned his face, barely anyone talked, the burnt and hot cane, exuding bubbles of boiling juice, when grabbed it would stick to the hand, he did not know whether he was hungry or thirsty. . . .[5]

> pronto el machete lo empezó a lastimar, se orinó las manos y eso evitó que se le levantaran ampollas, entrado el mediodía la temperatura era insoportable, el aire caliente hasta le quemaba la cara, casi nadie hablaba, todos entrados con el machete, la caña quemada, caliente, sacaba burbujas de jugo que ya era miel hirviendo, al agarrarla se pegaba en la mano, no sabía si tenía sed o hambre. . . . (18)

> nima pejki ki makokoua i tepos, mo ma axixki ua ijkino ayok ma apolonki, kema mo echkauiyaya tlajko tonal tlatotonili ayok mo ijiyouiyaya, ajakatl totonik, ok kin ixtlatiyaya, amo chene aka tlajtoyaya, ouatlachichinoli ua totonik, ki kixtiyaya i ayo posontok tlen mo nekchijtojka, kema ki itskiyayaj tlatskiyaya in mako, amo ki matiyaya intla mayana oke amiki. . . . (19)

The passage, with its cadence of short, repetitive phrases, speaks to the hazy, feverish state of working in the heat. Mundo's body is burned by the blazing sun, and the scalding hot liquid oozing from the sugarcane stalks. The conditions to which Mundo is exposed sever his connection to his own body; the physical sensations he experiences become illegible, as he cannot even tell what would soothe his discomfort. This first scene of work on a plantation foreshadows the process we witness throughout the novel of how the exploitation of land damages the ecosystems yet also people's relationships to themselves and each other. Donna Haraway argues that labor is the basis of how we understand ourselves and the world around us: "Neither our personal bodies nor our social bodies may be seen as natural, in the sense of existing outside the self-creating process called human labour. What we experience and theorize as nature and as culture are transformed by our work. All we touch and therefore know, including our organic and our social bodies, is made possible for us through labour" (10). A violent relationship to the land and the workers that the plantations establish translates into a disconnection from one's own body and the surrounding human and nonhuman agents.

The last time Mundo and Nabor take up a job on a plantation intensifies the tendencies that are marked from the start. Not only do the bodily sensations become more intense, but also the relation to the land on which they work is even more antagonistic. The last chapter of the novel abounds in descriptions of how the jungle where they work is being destroyed, as plants are being eradicated and animals killed. In other words, this job stands out as particularly violent. Moreover, Mundo and Nabor's motivation for taking it up is different. This time they are in no financial need of an extra job; they just want to escape their home village, that is, what used to be the place they longed for and hoped to improve by taking up the seasonal jobs. The incident described as the impulse for the two friends to leave is Nabor's argument with his wife. Drunk and out of control, Nabor takes out his anger on his family, alleging that the fault is his wife's, as the improved economic status has made her too arrogant. While critical of Nabor's lack of self-control, Mundo is sympathetic to his discontent with his wife, and they decide to leave once more, to temporarily escape the family conflicts.

Misogyny becomes a central element of the last part of the novel. Arias notes that the third chapter constitutes a significant shift in the narration: "It turns into a text about the masculinist complicity in going after adventures of a sexual kind" [Se convierte más bien en un texto acerca de la complicidad masculinista en la persecución de aventuras de índole sexual] (166). Gender oppression comes to the foreground, as women are mistreated and simultaneously perceived as a peril. Off-handedly, going through different memories from the many jobs he took up, Mundo recalls how a woman pregnant with his child asked for support from him, and he managed to escape, leaving her alone on a bus. When on their last job, both friends look for sexual and romantic relationships with the women they encounter, while dismissing women for their sexual activity. Their way of demonizing sex and women reaches its peak in their contemptuous treatment of sex workers. In some ways, we could perceive the novel as confirming those discriminatory ideas; in the end, Nabor dies from a venereal disease he contracts from a sex worker.

Both Nabor and Mundo go against the values they previously held dear as they continue engaging with the plantations, yet Mundo successfully covers up his different transgressions, while Nabor explicitly breaks the structures of the community, most of all by bringing the violent behavior back into their home village. Given that, Arias reads the ending of the novel as the reconstruction of the values of the community: Nabor has to die for the values to be reinstated (170). Yet the text reads as less reparative if we consider the death of Nabor as a sign of the damage done to the human and nonhuman world by colonial and capitalist practices. The name of the protagonist, after all, is Mundo, which is Spanish for "world." The effects of the transformation of labor practices and treatment of land seem necessarily broader than Nabor's individual faults and cannot be contained in a death or kept at bay by perpetuating the violence somewhere else rather than a place considered one's own.[6]

During Nabor and Mundo's last plantation job, the workers observe a group of monkeys that come up close to their encampment. Nabor captures and ties up one of them in a gratuitous act of cruelty. Other monkeys come to the captive's rescue and manage to set him free. On the one hand, Nabor's actions signify the hostile shift in the relationship to the nonhuman world that the workers experience. On the other, the scene speaks to a fantasy of the natural world, of a lost natural order. The monkeys are described in a highly anthropomorphic manner in their appearance and their movements. Rather than observe or describe them, the novel projects onto the monkeys what it desires for a human community. Haraway asserts that science has long treated monkeys as "natural objects unobscured by culture," offering a glimpse into the natural character of human beings before we were changed by culture (14). Monkeys are taken as "the base of supposedly natural, integrated community for humanity," Haraway finds, while in the analysis of their relationships, multiple biases cloud the possibility of observation, such as assuming the dominance of particular individuals as natural (19).

Although this scene might seem like a way to value animals, it shows, rather, human self-involvement, which limits greatly the capacity for insight. In this case, telling natural histories becomes an effort explicitly centered on the human while at the same time, through this distortion of perspective, hinting at how the exploitation of the nonhuman world makes people less capable of observing and analyzing it. The last chapter of the novel is also where the text turns fully to a first-person narrator, soon after Mundo and Nabor arrive at the last plantation job. While before it seemed like the novel played with the possibility of a plurality of perspectives, in the end it narrows down to just the individual, separated from the community and from the broader nonhuman world. Violent ways of interacting with the world seem to make us less capable of speaking of it, limiting our perspective.

The scene in which we see a more verisimilar observation of human interactions with animals is even more violent; it is the moment when one of the workers kills a crocodile. The man is armed with a chainsaw, and first he severs the animal's tail, and when the crocodile keeps on fighting, he continues to cut off other parts of its body.

> We saw as he cut his arms, and gave many blows to his head until one cut his throat, the animal soaked the man with his blood, when we saw him stop we drew closer, the animal did not have a head, nor arms, nor a tail, it still threw itself around, wanting to defend itself.
>
> Vimos cómo le trozó los brazos, y varios golpes con la motosierra en la cabeza, hasta que en una de tantas le trozó el cuello, él estaba bañado con sangre del animal, cuando vimos que quedó parado nos acercamos, el animal casi sin cabeza ni brazos ni cola, aún se aventaba, tratando de defenderse. (106)
>
> Tik itakej kenijki ki matsontejki, ua miyakpa ki kuatejtsonki ua san sejpa ki kechtsontejki, ya mo altijto ika netlapiali i eso, kema tik itakej mo ketski ti mo echkauijkej, ne tlapiali amo ki piayaya i tsonteko, yo i ajkolua, yo i kuitlapil, ua noja mo majkauayaya, ki nekiyaya mo manauis. (107)

From the start, the imbalance of power is clear. As the animal keeps on fighting, desperately, all it can achieve is to cover the man in its blood. This is a minor way of persisting, of leaving its mark on the world.

In this image, storytelling can be seen as driven by the desire to observe, gather, and preserve evidence of lives and matter. In other parts of this chapter, we see how faulty stories are told instead. They obfuscate rather than explain, as the mortal danger comes from women and sex, and animals cannot be really seen beyond what we see of ourselves in them. But then moments such as the killing of the crocodile arrest us, forcing us to look rather than project, to observe and describe rather than use. Instead of a domesticated anthropomorphism or a taxidermized lifelessness of an animal, we are faced with its slaughter, standing out from the novel for its graphic violence, seemingly gratuitous. Such moments are deeply unclear. After all, how can we read bloodstains? They manage, however, to interrupt human narration, leaving space for what refuses to be incorporated into hegemonic versions of natural history.

*

The opening of Fajardo-Anstine's *Woman of Light* is saturated by nature. Each sentence is threaded from natural elements making an appeal to our senses. Many scales of the natural world meet in these opening scenes: "the sky was so filled with stars it seemed they hummed," while a baby is left by a riverbank "turkey down wrapped around his body, a bear claw fastened to his chest" (xvii). Stars, furs, crickets, cedar, thistle, ice, and human bodies all come together. They construct images slowly, with attention to their multisensorial details. It seems like many elements need to work together for each paragraph to turn into the next. The actions of people are slowed down, as if caught in intricate oil paintings; they do not seem to take priority over the organic and inorganic world that they blend into rather than stand out from. Otherwise, the actions might seem quite dramatic: one woman leaves a baby by a river, and another finds him right in time before he freezes. It is by reading the signs within the surrounding world that Desiderya interprets the situation of the baby she comes upon. She looks into the reflection of the sky in the water on the baby's cheek to realize that he was "left to be found" (xix).

After the opening scenes of the novel, this intensity of the elements of the natural world subsides little by little, leaving more room for the story of the humans inserted into this space. The baby found by the river is Pidro, the grandfather of Luz and Diego, the youngest generation of the family we meet in the novel and the one that we get acquainted with best. The focus of the novel is on two years of their lives, spanning 1933–1934, while it offers occasional glimpses into the lives of their parents and grandparents. Luz and Diego live with their aunt, Maria Josie, in Denver. These are the times of the Great Depression and of civil unrest—the novel depicts and hints at ways in which workers and people of color organize to fight back against different forms of injustice they are facing. On the backdrop of this sociopolitical panorama, we see Luz's and Diego's personal lives unfold.

The precarious balance they live in is shaken up when one night Diego is brutally beaten up by a group of men and is forced to leave the city, fearing further violence. We learn the backstory later: when one of his lovers gets pregnant and wants an abortion, they seek out a woman who claims to be able to help. Yet the concoction does not work as expected, and as they leave, Eleanor Anne starts bleeding profusely. When three white men approach them, Diego is immediately accused of hurting Eleanor Anne and runs away, only to soon be found by his lover's relatives, who are the ones to attack him and make him flee from the city.

Many more details would be needed to accurately represent the stories of Luz and Diego as well as those of their relatives, friends, and ancestors. A through line, however, is how the novel tells different stories of love and affection inserted in a world saturated by injustice. Class, racial, ethnic, and gender discrimination, conjugated in different ways, can be read within all the personal narratives and interactions. Language participates in giving shape to this world: the generation of Luz and Diego's parents still speaks Tiwa besides Spanish and English; the almost exclusively English narration reflects the way in which not only Tiwa but also Spanish are pushed out of the lives of Luz and Diego, forbidden in school, and looked down on in public contexts. Nonhegemonic languages are driven out of the public sphere and out of the intimate spaces of private conversations and storytelling, ultimately from one's own mouth.

The modes of care and affection available through storytelling are reduced by oppressive language politics. Stories of the human and nonhuman world are in various ways erased from or distorted within the hegemonic texts and institutions. Indigenous people were not considered authors and audiences of novels, and some languages were deemed unfit for public life. Similarly, the way in which archival materials are collected reflects a series of biases regarding which stories deserve to be kept and retold. Rigoberto González relates that these exclusions were an important motivation behind *Woman of Light*:

> [Fajardo-Anstine] sifted through the archives and records of places such as the Denver Public Library, the History Colorado Center, and the Center of Southwest Studies in Durango, Colorado, but came up short. "We were not in the archives," she says with chagrin. "The historical record was filled with stories of white Americans. I even came across a collection of Klan hoods with family names sewed inside the brim. But my people's narratives were few and far between. So I went to the storytellers, the elders, to get their accounts of history."

Shelli Rottschafer points out that Fajardo-Anstine's characters often go through a similar process as they "reconstruct a cultural practice for themselves through traditional ecological knowledges" (2). Telling stories and listening to them are a form of care, while they also reflect the ways in which affection is distorted.

Arguably, Luz's and Diego's connections to the nonhuman world become figures of these forms of affection marked by the dynamics behind sociopolitical relations. The way we meet the protagonists has a lot to do with natural history discourses. We are introduced to them while they are making some extra money. That is, we see them busy with what is presented as their side gigs but also their passions and talents: Luz reads tea leaves, and Diego performs as a snake charmer. These occupations draw attention to the imperfection of our forms of care: they are vitally important yet pushed to the side, never becoming full-fledged professions; they are haunted by a possessive desire and a threat of violence. But they also become experiments with modes of enacting affection that would waive control. The reader can get a better sense of this dynamic in the scenes of Luz's and Diego's performances.

In one scene, Luz is reading tea leaves for a man worried about his health. She looks into the bottom of the cup, and "along the edges, she saw a pig's snout, and deeper into the mug, far into the future, she glimpsed a running wolf" (3–4). She interprets the images as a bad case of gout. Her visions, however, often become more intense than those symbolic creatures. Soon afterward her cousin, Lizette, comes by.

> Delicately, Luz placed the cup brim down, draining leftover tea on a cloth napkin, the brown water bleeding onto the fabric. She turned the cup counterclockwise three times before flipping it over and gazing inside, the leaves darkly drenched like ground liver. A star, a boot, stray images along the ridge. Luz focused until the symbols blurred, giving way to another view, a moment caught like a trout from a river. Black hair rising and falling over white sheets, Lizette's curls perfumed with rose water. A long, airy moan. Teeth against floral pillows, a curved toe hitting a metal bed frame. Luz closed her eyes, she turned away from the cup.
>
> "Wow, Lizette," Luz said, flatly. "You aren't even married yet." (8)

Luz's reading focuses our attention back on the natural world. Each element is intensified by its similarity to others: the tea leaves gain the thickness of ground liver, the water darkens the cloth like blood. Symbolic images soon become a vivid glimpse into Lizette's sexual life, which Luz catches expertly even though it is slippery like a fish. Similarly as Desiderya who could understand the situation of the baby she found in the woods based on his inscription onto his surroundings, Luz is able to read the world as a series of connected vessels and textures to get an astoundingly clear image of what she is inquiring. As if aware of how detailed the view becomes, Lizette gets angry at Luz and demands, "Stop looking!" (8).

Soon afterward, Eleanor Anne comes to Luz for a reading. After introducing herself, she says, "And you're Little Light?" to which Luz responds, "Only my brother calls me that" (10). The initial distrust Luz feels when approached by a white, English-speaking woman is strengthened by this usurpation of intimacy. This is also the first time readers get to know Luz's familial nickname, hinted at in the novel's title. This appropriation of a term of affection is followed by Eleanor Anne asking Luz where she is from, in a way experienced by Luz as intrusive and marking her as an outsider of the hegemonic community. Consequently, there seems to be some reluctance in her initial gaze into the leaves, but soon a complex vision surfaces: a little girl in a strange place performing a dangerous trick.

> Then something strange, off-putting, the tea leaves seeming to drift like a blizzard over golden plains until Luz saw a place she hadn't seen before, twilight, a grassland marked by a dirt road, a lighted caravan of horse-drawn wagons hobbling along the path. . . . A pack of white men and women had gathered, and the girl asked if they'd like to see a trick. She spoke English like a grown-up. "Like with cards?" asked a child among the audience, and the girl shook her head. She pulled a fire poker from a satchel. It was a foot long with a hawk's talon at one end and a spiral handle at the other. The girl plunged it down her small throat, turned the handle, and brought it out again, clean. "That's no trick," yelled a man from the crowd. "Better trick would be to gut yourself, gypsy." The pale mob laughed then, edging toward the girl as one. (12–13)

Luz gives into the leaves. She is drawn in by the shapes and forms into a vision she cannot control. The tea leaves take over her sight as well as other senses; in some visions she tastes blood, and sometimes she loses consciousness. Rather than imposing symbols onto the leaves, as in the more superficial looks into the teacups, the leaves give shape to vivid, multisensorial experiences. This puts nature in an active, nonsymbolic position within the narration. Elements of organic matter seemingly as puny as wet tea leaves can take over the senses of willing human subjects. The hegemonic hierarchy is shaken when telling the story of the future, past, and present passes through giving up ownership and control.

Luz is led into a disturbing and moving scene where a little girl is shown as both fragile and powerful. The crowd singles her out as ethnically different. The trick she performs is impressive and threatening. The danger of moving a blade in and out of her throat can be read as connected to storytelling; there is a treat in poignant words and stories as well as in the mouths and throats that house them. But these experiences do not always become narratives. While they appear in the novel, thus becoming part of the story for us as readers, often Luz refrains from telling them to others. We can see it in the scene of the reading for Eleanor Anne. Luz keeps this image—and all the possible stories it holds—to herself. Given that she performs the reading for Eleanor Anne, this image is co-created by them and could become a shared moment, a place of mutual care. Yet the beginning of their interaction does not allow this space to open, and instead, Luz keeps the reading short and banal. All she says is that she saw "some kind of circle" in which Eleanor Anne sees a ring and hopes, "Maybe I'll be married" (13).

Diego is announced to us before we see him perform: "'Arriba, mira, you'll see him soon. Diego Lopez,' the boy shouted. 'Snake Charmer'" (9). Luz's brother has quite a different job. Rather than seeking answers through the clues of the natural world, Diego can teach it to act according to his will. Right before the reading that Luz offers to Eleanor Anne, we see a snippet of Diego's performance.

> Diego called to them. His Reina. His Corporal. The snakes rose together in a braid, their brawny bodies held apart, creating a space where Diego's face could be seen in the gap, calm and unflinching, his eyes highlighted in black kohl and his mouth painted red. He reached for his snakes, lifting them higher by their fangs. The crowd roared as Diego released them from his grip and the reptiles fell to the floor, playing dead at their owner's feet. (10)

The tricks continue. In the passage, Diego's face is framed by the snakes, which fold to his words and gestures. He shows command of the stage and the animals. The passage repeats possessive pronouns, and when the snakes fall to Diego's feet pretending to be dead, he is called their owner, foreshadowing how later they will die by his hand. The crowd is pleased with this display of animals obeying a man: "Coins like hail clinked over the wooden stage. The entire world, even the glistening river and creek, darkened as Diego moved into his next trick" (10). The spectators are happy to see this feat of human domination, yet the rest of the world darkens at this performance as if tapping into the violence in which it is rooted.

In other scenes, Diego takes care of and trains the two snakes, Reina and Corporal. We see how he saves Reina when she is thrown into the river to drown and how he tends to both snakes in a loving and joyful way. Yet the flipside of this care is the violence of ownership and control. When having to flee Denver, he cannot take the snakes with him easily and resolves to kill them instead. Luz opposes his decision; she would rather he leave them with her. Diego's final argument is concise and stern: "They're mine, Little Light" (46), he says to his sister. He sees his care for and affection toward the snakes as a basis for ownership, most fully experienced in having the power to either save or kill them.

At the end of the novel, Diego comes back to Denver. He has a new snake with him, one he calls Sirena, the nickname he uses when he leaves bleeding Eleanor Anne after she tries to get an abortion: "I'm so sorry, mi sirena" (148). Later we see Eleanor Anne in a convent asylum and learn that she did not manage to terminate the pregnancy. Diego's family figures it out too, and upon his return they claim the baby from the orphanage where she was placed. In some ways, this scene could be read as reparative. Diego and his baby are brought back into the family fold; in the last scene of the novel Diego asks Luz to tell his daughter all her stories, thus creating a sense of circularity of the novel. In that sense, the double homecoming that closes the text is a hopeful event. But we can also read it as another example of broken forms of care. The snake, Sirena, that Diego comes back with repeats the model of affection based on ownership. Sirena substitutes the two snakes he killed, as if the animals were interchangeable. Moreover, given the snake's name, in a way it stands in for Eleanor Anne, about whose fate we learn nothing more. Diego's return with Sirena highlights the violence undermining the affection he offers others.

Luz's relationship to the organic matter she looks into is different. She sees stories in tea leaves but cannot really tell them; they escape her. They are based on someone else's fate, yet they take shape in her eyes and mind, mixing with her memories and visions. They are neither hers nor of anyone she tells them to. These are unclear stories, rooted in paying attention to minute details of nature. They mix temporalities in a way that does not respond to the human experience of time. And they create sensorial feelings that do not obey the logics of causality. The stories seen in tea leaves are moments of opacity, when observing and narrating exceeds linear and rational structures. The tea readings reclaim a broader set of narrative practices, showing fortunetelling as a mode of paying attention to the world in ways that are not sanctioned by the everyday concepts of logic and reason.

The strong effects that those stories have on Luz show that giving in to them can be risky. This form of storytelling is presented as a dangerous endeavor, of leaving space rather than taking it and pushing oneself out of a personal perspective into a mode of narrating outside personal control. Those stories are only sometimes told, as Luz is unable or unwilling to give space to some of the images that open up to her. As the novel illuminates this difficulty, it keeps us looking for possible ways to make the practice of writing and reading novels closer to such transgressive, broadly communal practices, in which voices would join and follow different guides, including animals, plants, and inanimate matter.

*

As I was finishing up this essay I made my way to the Harvard Natural History Museum, wondering whether having spent some time thinking about natural history would make this visit different from the previous ones. I headed first to the usual rooms that I never cease to enjoy. The famous glass flowers were fascinating as ever. The mineral gallery yet again displayed the casual glamour of rocks whose beauty seems just as obvious as mysterious. A lot has been written on how they formed as well as how over the centuries they have traveled and been admired and assigned value. I wondered whether I could learn to tune into their glimmers and shapes in new ways; ultimately, I moved on to other gallery rooms. As always, the intense gaze of the taxidermized animals made me quite uncomfortable.

But then I noticed something I had never paid attention to before: in the arthropod section, there was a separate exhibit in one of the corners: *The Rockefeller Beetles*. Rockefeller is a name that pops up often enough as that of the owner or founder of different institutions and initiatives. But that afternoon it stood out to me—the Rockefeller mines feature in the background of *Woman of Light* as part of the general sketch of the poverty and abuse of workers in the Denver area. The novel makes a veiled reference to the Ludlow Massacre of 1914, twenty years before the main plot. The massacre occurred during a strike in which mineworkers demanded, among other terms, safer work conditions and an eight-hour workday. The strike was violently suppressed by forces composed of the Colorado National Guard and privately hired militia who killed many protesting workers and their family members. In order to repair the family image after the massacre, John D. Rockefeller Jr. invested in public relations. The adopted strategy involved retelling the story not just of the Rockefeller family but also of the workers themselves. In Robin Henry's analysis, "As an adherent to the Progressive Era principles of welfare capitalism, the gospel of efficiency, and muscular Christianity, Junior relied on organizations like the Young Men's Christian Association (YMCA), longitudinal statistical studies on vice, the new pseudoscience of eugenics, and eventually the law in his attempt to reshape his workers into docile company men who embodied middle-class masculinity" (26–27). Another measure Henry describes was the "Americanization" of foreigners; altogether, those efforts were supposed to pull workers away from union organizing and socialism (30).

The beetle collection displayed in the museum was created by David Rockefeller, the grandson of John D. Rockefeller Sr., the first US billionaire, and the son of John D. Rockefeller Jr., responsible for the family company during the Ludlow Massacre. David Rockefeller was described as a "banker and philanthropist with the fabled family name who controlled Chase Manhattan bank for more than a decade and wielded vast influence around the world for even longer as he spread the gospel of American capitalism" (Kandell). At *The Rockefeller Beetles* exhibition, visitors learn that collecting beetles was something David Rockefeller started at a young age and continued throughout his life. Having traveled often and widely, he amassed a collection of 150,000 specimens that first were housed in one of his estates and then donated to the museum. On a gallery wall is posted a reflection by his daughter Eileen Rockefeller Growald. "Through beetles," she observes, "my father found his own sense of humility . . . he has kept his eye to the ground, to little things that in life really run the world." This collection relates to the broken forms of care depicted in these two novels, when the awe before nature and its variety translates into taking possession of the biggest possible slice of that admired world. Echoing the idea celebrated in the British Museum of collecting the world, it invites us to reflect on who collects, categorizes, gifts, and displays the stories of the "little things that in life really run the world."

We saw how the natural history discourse would treat some people, primarily Indigenous, as the objects of its study. Yet natural history is human history in another way: its stories about the nonhuman world speak of the humans that thread them. *The Rockefeller Beetles* exhibit is an example of how scientific discourse and museography are deployed to shape the memory of an individual or family. More broadly, natural history museums speak also of the discipline and its history. At times they explicitly present telling those stories as part of their mission. If we take another look at the Natural History Museum in London, we will see that it also displays a collection of portraits. In its catalog we read,

> A number of portraits were purchased, including those of Gould, Linnaeus and Owen, although the point was made on each occasion that "in the ordinary way the acquisition of portraits is no part of the function of this museum." . . . In the recent years it has become accepted that the acquisition of portraits and other natural history paintings is a proper use of the Museum's resources, and that they form an important adjunct to the collections of specimens, books, and manuscripts. (Thackray x)

Thus, the museum is writing and collecting its own history. It displays portraits of multiple men and a handful of women: zoologists, geologists, collectors, taxidermists, authors of monographs and treatises. Among them are some whose contributions are not seen favorably today, like an entomologist about whom John Thackray writes that "much of his work is now seen to have been hasty and careless, and few of his species survive in the modern literature" (44).

El infierno del paraíso and *Woman of Light* foreground epistemological and narrative modes that are not part of hegemonic literary forms and scientific research. They reach into oral storytelling and give priority to materials and languages that are often absent from historical archives and literary markets. This process involves attempts to make space for the nonhuman world in ways that shake up the hierarchies of constructing meaning. By transforming the genres and discourses tied to modernity, these Indigenous writers ask how to think through new forms, curious about modes of care for the world that are not tied to dominance over it. At times, reading these novels is like walking around a museum where natural history is told and severed by its telling; there is a display, but filled with grief and sadness, not with a more triumphant desire to encompass and explain it. Other times, the novels offer unclear images, forcing us to pause before tea leaves and bloodstains and asking to what extent we can read them without imposing pre-established meanings onto them.

Georges Didi-Huberman reflects on how attempts to capture images change them forever, comparing them to butterflies pinned in displays: "We know well where to find the known butterflies: in the glass cabinets of our natural history museums. But we do not know, manifestly, where the living butterflies go and where they come from as we see them pass us by, the wandering butterflies. Suddenly they proliferate, suddenly they disappeared" [Nous savons bien où se trouvent les papillons connus: dans les vitrines de nos muséums d'histoire naturelle. Mais nous ne savons pas, à l'évidence, d'où viennent et où s'en vont les papillons vivants que nous voyons passer, les papillons errants. Ils prolifèrent soudain, soudain ils ont disparu] (15). In other words, we can thoroughly analyze that which we halt and hold, but this will be a correspondingly limited, static knowledge. *El infierno del paraíso* and *Woman of Light* invite us to leave space for the nonhuman in ways less structured than display cabinets, experimenting with how to learn from what we cannot grasp.

Notes

1 To give just one example, the Field Museum of Natural History in Chicago includes materials of the nonhuman world together with archaeological objects from around the world and with Indigenous objects across history. The exhibition of Indigenous artifacts has recently been reshaped, after years of criticism. The museum website explains, "*Native Truths: Our Voices, Our Stories* is a new permanent exhibition at the Field. It replaces and re-examines the previous Native North America Hall that existed in this space for many years, which was created without the input of Native people themselves." What these efforts do not remedy, however, is that Indigenous cultures are presented within a natural history museum, and this contextualization makes implicit claims of them pertaining to a different temporality and having a reduced sociopolitical, artistic, and intellectual importance.

2 Deloria contends, "Their material traces were commonly organized around three categories: American history, in which they made a quick appearance and then disappeared; anthropology, in which they illustrated social evolution or, at best, cultural relativism; or art, in which their objects were recontextualized around form more than function, and in which they served as a primitivist foil for American and European modernism" (109).
3 Most of all there has been an important movement to remove human remains and sacred objects from museum exhibitions and storage spaces. Within the United States, the removals were legislated through the Native American Graves Protection and Repatriation Act of 1990.
4 Unless otherwise indicated, translations are my own.
5 Since the Spanish and Nahuatl versions of the novel do not coincide exactly, my translations take into consideration both versions.
6 Arias observes that in the last chapter of the novel, no geographic markers or names of plants appear, but the descriptions of the flora and fauna allow us to identify the place as Chiapas, the site of important Indigenous revolutionary movements. The sense of pride and triumph, however, is ruined as the jungle is damaged by the exploitation of its resources, a reference to the actual exploitation of the forests in Chiapas, in the name of modernization (166–167). Not mentioning names of places and plants in this final chapter, the novel laments the loss of their meaningfulness in the eyes of the narrator.

Works Cited

Amador Ramírez, Crispín. *El infierno del paraíso.* Instituto Mexiquense de Cultura, 2005.

Arias, Arturo. "Nahuahtlizando la novelística: De infiernos, paraísos y rupturas de estereotipos en las prácticas discursivas decoloniales." *Revista Ciencia y Cultura* 17, no. 31 (2013): 153–179.

British Museum. "The British Museum Story." https://www.britishmuseum.org/about-us/british-museum-story.

Chacón, Gloria Elizabeth. *Indigenous Cosmolectics: Kab'awil and the Making of Maya and Zapotec Literatures.* University of North Carolina Press, 2018.

Deloria, Philip J. "The New World of the Indigenous Museum." *Daedalus* 147, no. 2 (2018): 106–115. https://doi.org/10.1162/DAED_a_00494.

Didi-Huberman, Georges. *Phalènes: Essais sur l'apparition, 2.* Éditions de Minuit, 2013.

Fajardo-Anstine, Kali. *Woman of Light.* One World, 2022.

González, Rigoberto. "Keeping the Stories: A Profile of Kali Fajardo-Anstine." *Poets and Writers* 50, no. 4 (2022). https://www.pw.org/content/keeping_the_stories_a_profile_of_kali_fajardoanstine.

Haraway, Donna J. *Simians, Cyborgs, and Women: The Reinvention of Nature.* Routledge, 1991.

Henry, Robin. "'In Our Image, According to Our Likeness': John D. Rockefeller, Jr. and Reconstructing Manhood in Post-Ludlow Colorado." *Journal of the Gilded Age and Progressive Era* 16, no. 1 (2017): 24–43. https://doi.org/10.1017/S153778141600044X.

Kandell, Jonathan. "David Rockefeller, Philanthropist and Head of Chase Manhattan, Dies at 101." *New York Times,* March 20, 2017. https://www.nytimes.com/2017/03/20/business/david-rockefeller-dead-chase-manhattan-banker.html.

Rottschafer, Shelli. "Land Acknowledgement: Surviving Displacement through Reclamation of Querencia in Kali Fajardo-Anstine's Short Stories 'Sugar Babies' and 'Ghost Sickness' Published in her Collection *Sabrina & Corina* (2019)." *Studies in American Indian Literatures* 34, no. 3/4 (2022): 1–26.

Siskind, Mariano. *Cosmopolitan Desires: Global Modernity and World Literature in Latin America.* Northwestern University Press, 2014.

Thackray, John. *A Catalogue of Portraits, Paintings, and Sculpture at the Natural History Museum, London.* Natural History Museum, 1995.

Thackray, John, and Bob Press. *The Natural History Museum: Nature's Treasurehouse.* Natural History Museum, 2001.

III

Ecocriticism, New Materialism, Posthumanism

7

The Necrospace of the Anthropocene

An Anachronistic Archive

Gisela Heffes

1. The Anthropocene

The term "Anthropocene"—derived from the Greek words *anthropos,* meaning human, and *kainos,* suggesting new—was introduced by Dutch chemist Paul Crutzen, who received the Nobel Prize in chemistry in 1995. It refers to a new geological epoch defined by the substantial impact of human activities on the Earth.[1] The effects and consequences of degradation and destruction are inscribed within the materiality of the planet, where the spatial and temporal scalar dimensions of human interventions shape, to some degree, the deep time of geological strata. The geological layers of the temporal breadth are vast and unfathomable, and they significantly exceed the human scale and the span of human history. But what does it mean for humankind to become a geological force? And what effect does it have on our lives to know that we are agents of change that alter the strata, the layers that account for formations and juxtapositions? Furthermore, is it possible to visualize and, *stricto sensu,* represent what, in light of its magnitude, escapes or exceeds our cognitive and perceptual capacities?

One of the most troubling aspects of the Anthropocene is the novelty of knowing that our present is accompanied by deep pasts and deep futures (Farrier 6). Stephen Jay Gould explains in *Time's Arrow, Time's Cycles* (1987) that the idea of geological deep time is so alien to us "that we can really only comprehend it as a metaphor" (3). If the Anthropocene exposes human agency on the vast scales that shape Earth systems and the life forms that support them, at the same time it makes us aware that humankind as a geological force is minimal when confronted with what Marcia Bjornerud describes as "wrinkled time," the "feat of compression" by which "Earth effectively wrinkles time—accordioning eons and juxtaposing moments in its long history—through the medium of rocks" (2024). On the other hand, deep (and geological) time acquires a dazzling and disconcerting aspect at once as an everyday element that accompanies and determines human and nonhuman life. From our continued dependence on fossil fuels or hydrocarbons to that of certain minerals as well as synthetic materials—including plastic, nylon, polyester, artificial preservatives, neoprene, polyethylene, and PVC—all of these components intimately connect us to a distant and prehuman past but also with a future (perhaps less distant) and the posthuman.

2. An Anachronistic Archive

If the Anthropocene as an epoch (or event) exposes an "effect"—a human geological force capable of altering the strata, the layers that account for formations, superimpositions, and sedimentations—it also reveals a "process" emblematic of terrestrial history, characterized by accumulation, absorption, discarding, mixing, and assembling, leading ultimately to the integration and elimination of various elements.[2] In other words, an epoch (or event) that, through its effect and process, functions as a material archive that organizes, categorizes, and inventories layers, mapping the sequence and structure of geological and temporal strata. This archive, whose sedimentations narrate stories of change, beginnings and endings, methods and transitions, and of temporalities inscribed as hieroglyphics on the rocks, operates on a different scale outside the framework of modern, teleological rationality. It therefore serves as an anachronistic archive, functioning not only against the flow of time but also as a discrepancy within the chronology of history, thereby challenging its linear sequence. One of the definitions of "anachronistic" is an error in chronology. According to Merriam-Webster, "anachronism" refers to a person or thing that is out of its chronological context or place; specifically, something from a previous era that appears incongruous in the present.[3] An anachronism is therefore outside of its time but understanding time, in turn, as a sequence and a consecutive ordering.[4] The error, then, should be understood not merely as a mistake but as an interstice through which to postulate a counterhistory, an unofficial history, an alternative genealogy to configure other archives, and along with them, to devise other forms of knowledge shaped by the archive's silences and erasures. I depart from the idea of anachronism postulated by Georges Didi-Huberman in *Devant le temps* (*Ante el tiempo* 154) as a model for interrogating history and in connection with Walter Benjamin's dialectical paradigm and notion of "discontinuities" and "anachronisms of time." In doing so, I explore how the geological layers of the Anthropocene can be interpreted as archives, a phenomenon Bjornerud refers to as "Earth's crustal archives." Furthermore, I reflect on the concept of the archive through two theoretical frameworks, the first being Achille Mbembe's notion of necropolitics and necropower, and the second being Walter Benjamin's idea of collecting, specifically that which possesses value beyond commodification and has the capacity to preserve remnants, detritus, and other elements teetering on the brink of extinction.[5] This seemingly contingent archive serves as a sovereign expression, as Mbembe highlights in his analysis of necropower, that regulates mortality by determining who lives and who dies according to a definition of "life" closely tied to the exercise of power ("Necropolitics" 11–12). In a broader sense, the anthropogenic necrospace, shaped by both organic and inorganic materials and forms, will gradually and inevitably give rise to anachronistic archives.

3. The Necrospace of the Anthropocene

An archive in Benjaminian terms supposes a noninstitutional archive; the official meaning, originating from Greek and Latin, derives from town hall or municipality and government offices, and refers, in turn, to ideas of beginnings, origins, and domain. Order, efficiency, completeness, and objectivity establish the principles of archival work. On the contrary, the archive for Benjamin consists of "images, texts, signs, things that one can see and touch, but that also constitute a reservoir of experiences, ideas, hopes, and that can be analyzed and cataloged by the inventory manager" (Wizisla 11). This is exemplified in "Edward Fuchs, Collector and Historian," where Benjamin describes the collector as someone who not only acquires objects but also transforms them, endowing them with connoisseurship value rather than utilitarian worth (Benjamin 3: 260–302). The necrospace of the Anthropocene as an anachronistic archive implies, therefore, collecting the tangible and intangible, "useless" debris that modernity has discarded: human and nonhuman bodies, objects consumed and discarded, gas emissions that come back in various forms, whether drought or flood, wildfires or hurricanes, or at scales and frequencies unimaginable in other times.

To illustrate how this archive unfolds and evolves over time—compressing and expanding the processes of erosion, sedimentation, and the formation of the necrospace—I will analyze three aesthetic works from the early twentieth century to the early twenty-first century to highlight significant continuities amid disruptions. I am not, however, suggesting that these disruptions or fractures in the discursive continuum of modernity are fully addressed in this article; on the contrary, they evince patterns of repetition that challenge the alleged seamless narrative of progress and economic, political, and cultural development. The three works under consideration are the short stories "Los pescadores de vigas" (The Log-Fishermen, 1913) by Uruguayan writer Horacio Quiroga and "Agua" (Water, 1935) by Peruvian writer José María Arguedas, along with the novel *El Rey del Agua* (The King of Water, 2016) by Argentine writer Claudia Aboaf. This novel is the second part of the trilogy that begins with *Pichonas* (2014) and concludes with *El ojo y la flor* (2019). In these narratives, water, whether abundant or scarce, reveals the crevices and frictions within an archive as bodies of water emerge as necrospaces. These literary texts trace genealogies that, when interpreted critically, shed light on the semantics of aesthetics, value, and power at the negligible and vast levels. Hence, I propose a fluid archive, specifically a material substance such as water, as a repository for understanding processes of inclusion and exclusion, of voicing and silencing, and of the construction of histories whose prevailing narratives have simultaneously intertwined with other stories from the margins. On the edges or periphery of these constructs lie, thus, the omissions and erasures of an equivocal history, the errors that claim a place from the interstices silenced by a hegemonic chronology.

It is significant to note that in these three examples, water serves as a catalyst for destabilizing the established order, suggesting through its fluid semantics a network of tributaries that diverge and meander in various directions contrary to the dominant flow. Yet, this fluid archive is not intended to be exhaustive. Instead, it seeks to offer a counterintuitive interpretation to underscore the elasticity, resilience, and freedom inherent in an anachronistic archive. It aims to evoke, as if such an archive could yield unexpected revelations, that the stories assembled from the margins form a tapestry of connections and kinships whose riches—differing from those extracted to depletion—remain latent, like a vital force waiting to narrate its own story and shed new light on the present.[6]

4. The Space of Writing and the Writing of Space

The deep time of the Earth shapes our present through geological strata and evolutionary biodiversity as well as textures, processes, devices, and artifacts, all of which inform our experience and perception of/in modernity. Writing, for its part, expands its own scalar capacity when it connects human time with the deep pasts and futures of the nonhuman. The writing practice, in its own way, evokes a geological effort as it uncovers traces, remains, relics, and material vestiges from various time periods to blend the archaic with the contemporary, the microscopic with the planetary, the local with the global, as well as the past, present, and future. Similar to Benjamin's archive, which consists of textual fragments (such as citations, notes, and aphorisms) that document the remnants of capitalist modernity and the discarded and forgotten, exposing the ruins of progress and showing that carrying the past into the present makes the present aware of its precarity, the writing practice rescues and breathes life to all kinds of residues abandoned by commodification.[7]

In Latin America, literature, cinema, and art in general offer a platform for experimentation, contestation, and in some cases, contention that shifts the attention toward these contemporary emergencies (by "emergencies," I mean their double meaning of "emerging" and "urgency"), establishing productive dialogues with new modes of critical and cultural inquiry, especially those concerning post-anthropocentric theories. Literary critic David Farrier proposes, in *Anthropocene Poetics* (2019), various aesthetic explorations to reflect on the scale of the Anthropocene and to visualize how, in this new geological epoch, the distant past and near future flow through the present in ways that are sometimes extraordinary (1–2). They surface as poetic assemblages whose materiality extends over time, uprooting and making their connection to their origins visible. This materiality, which amplifies the presence of human and nonhuman traces in the form of anthropogenic landscapes, fosters a divergent materiality that sometimes aligns with disrupted ecosystems.

The notion of space is central to the discussions concerning the Anthropocene. I draw on a theoretical framework specific to the environmental humanities, one that fluctuates between ecocriticism (attention to the natural and/or nonhuman environment), new materialism or neomaterialism (the agency of matter, also for many critics, post-anthropocentrism); and posthumanism (the deconstruction of human exceptionality) (Fornoff and Heffes 7). In particular, the publication of postcolonial ecocritical works interrogates positions and debates related to European projects of conquest, colonization, and global domination. Similarly, Mary Louise Pratt's definition of the "planetary imaginary" examines the futurity crisis accompanying the new millennium. This crisis is linked to environmental catastrophes and the new forms of thinking it has catalyzed. Many of these works foster a reevaluation of the imperialist and racial ideology on which many critical experiences, aesthetic imaginaries, and local socioeconomic conditions depended and still depend. However, Pratt's understanding of the crisis not only challenges a simplistic postcolonial perspective but also takes a significant step forward by advocating for the decolonization of postcolonialism: "Postcolonial inquiry bracketed out the so-called first wave of European imperial expansion, that is, the Spanish, Portuguese, British, and French colonial enterprises of the fifteenth through eighteenth centuries, all over the planet but most conspicuously in the Americas" (16).

Along with the question of an exploitable nature, other inquiries are articulated, such as the conditions under which vernacular and Indigenous communities were brutally forced to abandon their homes, their lands, their values, their natural and ecological visions, and their cosmologies to adopt those imposed by the culture of the West. While I generally agree with Pratt's position, it is important to recognize that a postcolonial ecocritical perspective introduces a valuable new line of inquiry by considering both landscape and seascape as active participants and actors in the historical process rather than just spectators of a unique human experience. Rather than a mere extension of postcolonial methodologies to the sphere of the material and human world, this modality requires accounting for the ways in which ecology works and is not always within the frame of human time and political interests (DeLoughrey and Handley 4). This stance provides a complex epistemology that recovers the alterity of history and nature without reducing one to the other and vice versa. Once Western practices of oppression and subjugation of human and nonhuman organisms took hold, the ecological and human impact became irreversible, as Graham Huggan and Helen Tiffin underline in *Postcolonial Ecocriticism: Literature, Animals, Environment* (2010). Against the current colonization project and as a liminal insurgency, it is urgent to promote an imaginary space made possible by the experience of place. In confronting the process of historical production in the Global South, postcolonial ecocriticism uses the concept of place to question temporal narratives of progress imposed by colonial powers. The regulation of bodies and their productivity, its relation to planning, and the appropriation and extraction of living and inert organisms in natural and constructed environments entails material interventions and interferences whose remnants weave, as a survival or postlife, thresholds of critique (Didi-Huberman 73).[8] Therefore, it is essential (and urgent) to inquire how aesthetic practices and cultural studies have defined and continue to shape our understanding of what is human and nonhuman.

5. The Time of Modernity

Against the teleology of modernity, the notion of place prevails as permanence—what is rooted—and as a mark of ahistoricity. Time, in its reverse, asserts itself as the fleeting and the ephemeral. I would like to depict this idea with an example: in his brief essay "Circling Eloh: A Meditation," Cherokee-born writer Lucien Darjeun Meadows draws on Cherokee rhetoric to reference what Cherokee scholar Rose Gubele defines "elohlogy," a term that derives from *eloh,* the Cherokee word for land, religion, law, history, and culture. Gubele describes elohlogy as a rhetorical form that combines ceremonial language with discussions about land and stirs shared memories. Specifically, the convergence between language, memory, and Earth intertwines to generate a shared and collective spatial experience. Using figurative language such as metaphors and imagery to emphasize the destruction and reclamation of the environment, Meadows writes, "We are where we are, much more than we are when we are. There is no when without a where. There is no we without a here" ("Circling Eloh"). Eloh privileges place and space over time. The "here" referred to in the essay, the spatial anchorage of the collective experience and memory, can turn instead into a necrospace when the channels that irrigate vital flows—of life—decompose into inert, toxic, or polluting layers. During his meditation, Meadows evokes the memory of his grandfather, which he connects with the present, where the threat of a potential flood from one of the dams in West Virginia—which is paradoxically situated next to a coal plant and a carbon deposit—puts the health of children at an elementary school only 150 feet away at risk. Mirroring the queries that Meadows asks himself throughout his writing, the meditation revolves around those questions in a circular pattern that emphasizes their importance through repetition. Challenging a linear and teleological view of time, he instead encourages readers to understand and meditate on the place itself. What has this place seen? Who has it met? What have these relationships left behind? Meadows discusses specific places and the different stories they have facilitated, endured, and experienced. All these seemingly unrelated stories are connected in a cyclical manner. Phrases like "The flood's coming now what do you do?" are repeated throughout to demonstrate how these accounts, despite seeming unrelated, share the same enduring purpose. Through both formal and discursive experimentation, a circular itinerary is also performed, simultaneously raising questions during the walk while problematizing the externalities associated with modern modes of consumption and production. Meadows captures that moment of confusion and contestation that emerges when the subjects and communities that inhabit these lands must confront the spatial harmfulness that defines and identifies them, transforming them into one more "risk society" among the thousands that add to the list daily. "Risk societies," as defined and conceptualized by German sociologist Ulrich Beck in his 1986 eponymous book, stand out for their operational modality, one that consists of an asymmetric condition from which to configure and address the risks and insecurities induced and introduced by modernization. Confusion and contestation because, as Stacy Alaimo suggests in *Bodily Natures* (2010), they require, in many cases, a scientific mediation that is not always accessible.

6. Topographies of Repression and Cruelty

In his famous essay on necropolitics, Cameroonian historian and political theorist Achille Mbembe suggests that the notion of biopower is insufficient to account for contemporary forms of the subjugation of life by power over death ("Necropolitics" 12, 39). The emergence of discursivities focused on the politics of life affects, Rosi Braidotti suggests, not only the practices related to death but also new ways of dying. Returning to Mbembe's conceptual proposal, Braidotti suggests that biopower and necropolitics are two sides of the same coin (2). For Mbembe, "The ultimate expression of sovereignty resides in power and the ability to dictate who can live and who must die;" exercising sovereignty is exercising control "over mortality and defining life as a manifestation of that power" ("Necropolitics" 11–12).[9] For this essay, I am particularly interested in the conceptual formulation that Mbembe develops to describe the topographies of repression and cruelty that emerge in conditions of coloniality and necropower, erasing dividing lines and confusing the modes of reaction and action. "Necro" is the suffix that in Greek refers to corpse or death. Necrospace would be, in this sense, the space that houses the corpses and death or both. It would be the result of a topography of repression and cruelty, whose organization and spatial layout respond to a regime of necropower that puts its sovereignty into operation to determine who/what will live or who and what should die. The material substrata of these topographies, therefore, are imbricated with mortality. In its sediments resides the hecatomb of modernity. If I transfer the notion of who to what, it is not so much to desubjectify the proposal of necropolitics or to subjectify the idea of space in the context of the Anthropocene, but rather to think, as Bruno Latour suggests, of ways to "distribute the agency as far and in as differentiated a way as possible, until we have completely lost any relationship between these two concepts of object and subject, which are no longer of interest, except patrimonial" (2). In short, to complicate its scope but above all to disrupt its limitations.

7. Aboaf, Arguedas, Quiroga

In the novel *El Rey del Agua*, the necrospace is defined as a fluid terrain, where the organic and inorganic, the garbage discarded for consumption, the toxicity coming from industries, tailings, and massive crops, along with the bodily remains of the disappeared from the last Argentine military dictatorship, are assembled, giving rise to nonspecific, liminal, and unstable materiality. The necrospace, thus, rebuffs essentialist and romantic notions of the natural world, in this case, water. Like *Pichonas* (2014) and *El ojo y la flor* (2019), the novel revolves around the sisters Juana and Andrea, whose father disappeared during the Argentine military dictatorship of 1976–1983. However, the novel also tells a story about the scarcity of water, its exploitation and privatization, and the spatial layout of a hydric topography in which human remains also navigate. In what we could tentatively define as an aquatic narrative, water acquires a necrospatial dimension involving the denaturalization of the fluvial networks surrounding the island of Tigre on the outskirts of Buenos Aires. The plot unfolds in the so-called Liquid Territory, a Tigre Delta narratively framed in the near future, where the sale of "raw water" and "river cemeteries" consolidates the King of Water's business. In fact, the taxonomies of water in the novel align with various aesthetic, formal, and thematic categories; for example, Andrea's sister, Juana, is immersed in complex networks and virtual environments, and this ongoing immersion deepens both the metaphorical and literal connection.

This semantic plethora, which I propose to read from the Anthropocene, has early precedents. Jason M. Kelly suggests that while most studies contend that the Anthropocene is a "measurable biophysical phenomenon," they often overlook that both the selection of key data points and the concept of the Anthropocene itself are not value-neutral: the Anthropocene often functions as a "metanarrative of modernity"—a narrative marked by energy and resource-intensive industrialization and capitalism, which have been accompanied by population surges, increasing flows of goods and people, the prominent role of nation-states, and calls for improvements in quality of life (Kelly 11). This narrative portrays a scenario in which humans have exploited the environment at unprecedented and ever-expanding rates, ultimately realizing that local actions have far-reaching consequences on a global scale (Kelly 11). The Anthropocene thus serves as a framework for critique, providing a way to articulate issues of excess, limits, thresholds, and boundaries.

For this fluid, anachronistic archive, I propose a new interpretation of Quiroga's story "Los pescadores de vigas" followed by Arguedas's "Agua," recognizing that both narratives are rooted in an emerging biophysical state manifested in the physical world as well as within our imaginations. In "Los pescadores de vigas," the Indigenous Candiyú acquires a phonograph from the Englishman Mister Hall in exchange for a few rosewood beams (Quiroga 139). This exquisite wood can only be obtained illegally when "the river really swells," which is when the logs come down from the cleared forests upriver (139). It is a profitable business for the Englishman, who owns more than one phonograph and whose transactions involve no risk or loss. The story is a strong critique of extraction policies. In the context of the export boom of the late nineteenth and early twentieth centuries, these policies revealed the presence of informal imperialism through various forms of neocolonization. Candiyú, who works in a yerba mate plantation of a presumably overseas origin, the Yerba Company, is dedicated to the illegal hunting of beams, which belong to foreign companies. After their extraction, the companies will send them abroad, where the centers of consumption and demand are located. Following a torrential rain that caused several treasured beams to collapse onto the riverbank, Candiyú dives in search of the prized beams to cover the cost of the phonograph Mister Hall had promised him in exchange. The hydric topography through which Candiyú flows gradually transforms into a necrospace, as his own body becomes entangled with inert organisms floating downstream at the same time: entire trees "ripped sheer from the earth," with "their black roots waving in the air, like octopi," dead cows and mules, along with "a good share of wild animals—drowned, shot or with an arrow still stuck in the belly," some jaguar, and "all the foam and floating lilies you like—to say nothing of the snakes, of course" (142). A drowned man with a slit throat metamorphoses the liquid necrospace into a multispecies ecosystem where various organisms—both living and dead—interact within a fluid territory, turning the waterscape into an anachronistic archive. Consisting of remains, errors, assassins and the assassinated, stories and counterstories, in the materiality of its tributaries, forms become blurred, textualities are superimposed, and truncated futures, origins, and temporalities are sedimented. Embedded within the framework of regionalism, Quiroga's text presents a significant contrast to the cosmopolitan and urban poetry linked with the expanding and rapidly modernizing capitals of the early twentieth century. It chronicles the environmental devastation caused by the hinterland export boom, encompassing extensive deforestation, droughts, floods, and soil erosion, and anticipates the forthcoming developmental policies that US President Harry S. Truman, in his 1949 inaugural address, announced when invoking a new program aimed at harnessing the benefits of scientific and industrial advancements for the enhancement and expansion of "underdeveloped" areas. This enrichment and growth would involve the adoption of new modes of production and extraction in Latin America, with costs becoming evident as a result still today.

8. Material Ecocriticism

The fluvial, fractured fabric of Quiroga's story evokes what Serenella Iovino and Serpil Oppermann refer to, in *Material Ecocriticism* (2014), as "storied matter." This reading views the material that shapes the world as a meaningful text resulting from the interaction of discursive and material forces, aiming to illustrate the dynamic interplay between human and nonhuman actors in shaping the world. In a later essay, Iovino contends that a material ecocritical perspective seeks to uncover a veiled clue embedded in its spatial textuality through its narrative expression, which acts as a prism capable of illuminating diverse stories—particularly those of nonhuman communities that inhabit a specific space, which is none other than the one containing us, as humans, and where our human lives are inserted (113-114).

If material studies (or new materialisms) have served as an analytical tool to critically inquire about the physical and natural environment, material ecocriticism in a broader sense derives from the idea that it is possible to merge interpretative practices with material expressions in such a way that the matter and meanings of global and mundane reality can be questioned (Iovino 114). This includes the activity of subatomic particles and exploring how the combination of toxic substances and practices engenders toxic spaces and bodies. Unlike the "ontological turn," the "material turn" aims to explore a new conceptual framework capable of theorizing the connections between matter and agency and understanding how bodies, nature, and meanings are interwoven. These can be cultural, social, political, and symbolic and can relate to the ways in which living matter organizes itself in self-regulating modalities. The interaction with matter is also necessary to retrace and recognize agential emergencies as they coalesce around social, scientific, and cognitive practices, along with how humans consume, exhaust, and reduce our material environment. That is why this analytical position regards matter as a text, a space for narrative and material stories (or storied matter), similar to a body palimpsest: extended bodies made up of matter and discourse in which stories are inscribed.

9. Fossils of the Future

What constitutes a fluvial, aqueous fabric of narrated—or storied—matter also operates as a space for contestation and dispute between power dynamics and the distribution of knowledge, geopolitical and economic systems, modes of resource extraction, and export, revealing the reduction and depletion of natural reserves. Additionally, it emphasizes how technology is often used as a means of benefit that is unbalanced and skewed. Twenty-two years after the release of Quiroga's "Los pescadores de vigas," in 1913, Arguedas published the short story "Agua," a narrative where water scarcity and, specifically, the unequal distribution of this vital resource, expose the tensions that characterize a colonial past that, throughout the story, extends and unfolds, making visible the connection with the present. The unequal distribution of a key resource, such as water, encapsulates the narrative within the broader framework of "uneven development," as delineated by geographer Neil Smith in his book of the same title, *Uneven Development: Nature, Capital, and the Production of Space* (1984). According to Smith, this concept illustrates the disparities in economic gains across different places by emphasizing how urban centers thrive at the expense of the hinterlands and how wealthy constructions coexist alongside impoverished neighborhoods. Furthermore, *Uneven Development* underscores the correlation between places that benefit unequally from the same economic processes.[10]

In "Agua," (post)colonial temporalities recur as a continuum in history: a pulse to remain that never ceases or disappears. If in Quiroga's story the fluvial territory manifests with all its fury and energy, in Arguedas's "Agua" it calls forth the scarcity of supplies. Not only, as Quiroga notes, "After a big drought, big rains" (141), but the same is true in reverse. Purposely refuting the implication of its title, "Agua" is all about dearth and paucity. Here, the hydraulic life-giving, essential source is practically inaccessible. The story, set in the small town of San Juan, takes place near an abandoned mine. As a vestige of an extractivist past rooted in both colonial and neocolonial history, the mine's presence reminds readers of the ongoing linkage between appropriation, exploitation, natural resources, and slavery-like labor conditions. Moreover, "Agua" composes a matrix that weaves the symbolic elements of storied matter to craft a material fabric from which the discursive draws relics and ruins of a past that endures. The necrospace where this fabric resides is a barren and scarce territory. The town, situated in the hills and surrounded by mountains, is the setting where the local and Indigenous populations gather every Sunday to demand water. As Pantacha explains to the narrator and protagonist, "Water, boy Ernesto, there is no water. San Juan is going to die because Don Braulio gives water to some and hates others" [Agua, niño Ernesto. No hay pues agua. San Juan se va a morir porque don Braulio hace dar agua a unos y a otros los odia] (Arguedas 10).[11] The *comuneros*' flimsy corn "almost does not move anymore, not even with the wind" [casi no se mueve ya ni con el viento], while that of Don Braulio "is fat, green, there is even mud on its soil" [está gordo, verdecito está, hasta barro hay en su suelo] (16), mud that only water can generate, because Don Braulio "was like the owner of San Juan" [era como dueño de San Juan] (27). In this desolate landscape, the community members "wept for water" [pedían agua lloriqueando] (21) and "if they didn't get a turn to fetch water, they left with all the bitterness in their hearts, thinking that their cornfields would dry up once and for all that week" [si no conseguían turno, se iban con todo el amargo en el corazón, pensando que sus maizalitos se secarían de una vez en esa semana] (21).

In the town of San Juan there are only "dry, brown, and leafless bushes" [los arbustos secos, pardos y sin hojas], a landscape of dehydrated rocks that is, to some extent, irreversibly hopeless (15). The "earthy slopes, the bare head of the mountains, the sand of the parched streams" [falderíos terrosos, la cabeza pelada de las montañas, la arena de los riachuelos resecos] (27-28) are embedded in the surroundings, giving the area a barren, livid appearance. Even though Braulio claims to own the town, the mountains, and, by extension, control the fate of community members who rely on the water he supplies at will based on his mood and preferences, Arguedas introduces two antagonistic perspectives: on the one hand is Braulio's caustic individualism; on the other are the *comuneros* (whose reference to community, "common," is evident) for whom the water, the mountains, have no owner: "Mama-allpa (mother earth) pours out water, the same for all" [Mama-allpa (madre tierra) bota agua, igual para todos] (25). More than anything, Arguedas crafts a story that offers a deep critique of the brutal exploitation of the Indigenous people—labor abuse, humiliation, and mistreatment that exposes the continuity of the past into the present—and also of the sociospatial distribution, organization, and transfer of power and hierarchies similar to those of the colonial period. Pantaleoncha, the old Indigenous water steward ("agüero," water master) who has long served the white landowning class (the "gamonales") by managing the distribution of water from the communal irrigation canal, expresses this categorically:

> As everywhere in Nazca, the *principales* also abuse the day laborers, continued Pantaleoncha.[12] They steal from men the work of the community members who come from the towns: San Juan, Chipau, Santiago, Wallawa. Six, eight months, they tie him up on the haciendas, withhold his wages; trembling with *tertiana,*[13] they put him in the cane fields, the cotton fields. Then they throw two or three soles coins in his face, like a big deal. Perhaps? Not even enough money for a remedy is given by the *principales.* On the way back, in Galeras-pampa, in Tullutaka, all along the road people spill out like little creatures, shivering; the Andamarkas, the Chillek'es, the Sondondinos die. They just stay there, with a pile of stones on their belly.
>
> Como en todas partes en Nazca también los principales abusan de los jornaleros—siguió Pantaleoncha—. Se roban de hombre el trabajo de los comuneros que van de los pueblos: San Juan, Chipau, Santiago, Wallawa. Seis, ocho meses, le amarran en las haciendas, le retienen sus jornales; temblando con terciana le meten en los cañaverales, a los algodonales. Después le tiran dos, tres soles a la cara, como gran cosa. ¿Acaso?

> Ni para remedio alcanzó la plata que dan los principales. De regreso, en Galeras-pampa, en Tullutaka, en todo el camino se derrama la gente; como criaturitas, tiritando, se mueren los andamarkas, los chillek'es, los sondondinos. Ahí nomás se quedan, con un montón de piedra sobre la barriga. (20)

The image is powerful. Men, like the inert rocks of the landscape, blend into the stony surroundings until all that remains of that inertia is dust, aridity, and mere drought. It is another necrospace where, in effect, the community members "are to die like dogs" [son para morir como perro] (20). Thirsty and dehydrated, the *comuneros* turn into decaying forms of life marked by a lack of water:

> Seeing the joy of the pampa, of the roads that go up and down from the village, the bitterness in my heart grew even more. There was no more Pantacha, there was no more Pascual, nor Wallpa; only Don Braulio was left. . . . Alone on that dry hill, that afternoon, I cried for the villagers, for their hungry little animals.

> Viendo lo alegre de la pampa, de los caminos que bajan y suben del pueblito, más todavía creció el amargo en mi corazón. Ya no había Pantacha, ya no había don Pascual, ni Wallpa; don Braulio nomás ya era. . . . Solito en ese morro seco, esa tarde, lloré por los comuneros, por sus animalitos hambrientos. (35)

Little by little, the *comuneros,* along with nonhuman animals, intermingle and become indistinguishable within another necrospace—a terrestrial, dusty realm: human residues that will remain scattered across the arid land until desiccation, forms of both animated and inert organisms that will be ingested by voracious scavengers and eventually form new geological layers, an anachronistic archive whose detritus generates an alternative history or a counterhistory in the furrows of its formations. What's more, in a scarcity-driven reversal, these abandoned bodies will become the fossilized remains of future human and nonhuman life. After all, "Agua" embodies a gloomy prognosis, warning us about a world where water—as well as other vital resources—are allocated disproportionately.

10. Intermission: Rulfo

The geological stratifications of the Anthropocene form layers that can be read as archives. These archives, apparently contingent, constitute the sovereign expression that Mbembe refers to when addressing the forms of necropower, namely, its ability to determine who can live and who must die and the practice of control over mortality based on a definition of "life" in correspondence with a manifestation of that power. When Arguedas writes "people spill out all along the road" [en todo el camino se derrama la gente], he is discussing or even anticipating another desolate territory, that of *Pedro Páramo* by Mexican writer Juan Rulfo (1955). Cristina Rivera Garza suggests in *Los muertos indóciles* (2013, *The Restless Dead,* 2020), "all the whispers," those that "go up or down the hill behind which the lights of Comala loom, stiff with cold" make up "the great necropolis populated by ex-dead" (10). Ex-dead, presumably because they were never buried. Or ex-dead because the machinery of developmentalism (the so-called green revolution) violently displaced and uprooted its communities, scattering their souls into barren and inhospitable places far away. Kerstin Oloff notes that the withered and depopulated landscapes in Rulfo's fiction—particularly in "Luvina," which was published only two years before *Pedro Páramo*—evoke the disastrous effects that agricultural and large-scale experiments had in Mexico. These experiments, conducted as part of the modernization process known as the Mexican Miracle, led to increased soil loss and subsequent desertification (Oloff 80). As a result, the materiality of the Anthropocene's anachronistic archive not only disrupts the postulates of modern progress—and the metanarrative of modernity—but is also erected from its residues: voices, bones, stones, and nails.

11. Utopia/Dystopia

Less than a hundred years after the publication of "Agua," water flow, essential for survival, reappears amid a new setback, this time from abundance. It is the water that, in official terms, is already privatized. This fantasy turned reality confronts the reader in Aboaf's *El Rey del Agua* as a fictional proposal that has already come true. Here, the new governor employs technology to purify the liquid treasure—or "blue gold"—and export it to countries where it is becoming scarce. Water is now private property, a monopoly with an owner, but instead of Don Braulio, it is owned by El Tempe Argentino (alias for "the King of Water"). While water is not yet scarce, it will be in the third and final novel that closes the trilogy (*El ojo y la flor*). Yet, in *El Rey del Agua,* water consists of a deposit where bodies and organisms rest, and the remnants of those disappeared during the military dictatorship dissolve and intermingle with organic and inorganic matter such as plastic, rusty metals, and garbage. Additionally, the water is drained for consumption. Overall, it composes a riverscape that, as Liliana Gómez and Lisa Blackmore note regarding the 2019 collapse of the

Brumadinho dam in the Brazilian state of Minas Gerais (2), harbors murky histories of capital flows, philosophical currents, aesthetic traditions, and lingering traumas that link spaces, times, and bodies in complex, meaningful, and unique ways. In *El Rey del Agua,* privatizing water means taking control of liquid resources but also privatizing memory, collective memory, and therefore history. While in Arguedas's "Agua" the communal and shared past is shaped by scarcity, discontent, dissent, and repression, the story does not necessarily succeed in changing the historical and economic conditions where it takes place. In Aboaf's novel, however, it is an abundance that, although unstable, is parceled out. Moreover, in segmenting, distributing, commodifying, and transforming raw material into the final product, human remains also circulate as consumable merchandise.

There is no need to imagine bottled water, land, and air. Like so many others, that unsettling imaginary has also crossed the borders of fantasies. A not-so-recent example is the 2012 film *The Lorax,* directed by Chris Renaud. As in the famous children's book by Dr. Seuss, the botanical life is artificial; there are no trees left, and unlike the story, the film incorporates an additional element absent from the text: the commercialization of air that the residents of this dystopian place rely on to survive. Not surprisingly, the air is polluted, and without trees, there is simply no oxygen to breathe. Seuss's account is categorical: extracting trees as a method of individual enrichment implies a grim future, a world at the mercy of those who control (and own?) the air (or water). The film, designed to sell promises and illusions to children eager for happy endings, ends with the final seed that opens the possibility of a generative future. There is hope. Not so in Aboaf's novel or the short stories by Rulfo, Arguedas, and Quiroga. To a certain extent, they are dystopian narratives, not so much because they adhere to the conventions of the genre but because they constitute necrospaces rooted in immediate and recognizable realities. The aesthetics of the Anthropocene blur the line between fiction and reality as well as between (u)topical and (dis)topical imaginaries. To some degree, the aesthetic practice of the Anthropocene feels realistic and close, conveying a sense of proximity. This may be because our universe, filled with both overwhelming abundance and scarcity plunges us (drowns us?) into a deep torrent, into the erratic waters of the predictable and the unrecognizable, of the known and the unknown. The intricate forms and poetics of emptiness and excess are articulated within this tension, forging a fiction of lavishness that seeks to downplay the urgency of environmental depletion. In essence, they enable necrospaces layered with both living and inorganic matter, gradually becoming a material repository, an anthropogenic archive.

12. Bank of (Bio)Data

In *El Rey del Agua*, the remnants of the murdered father flow through the liquid space, providing the latter with information, which is extracted to constitute an archive—another, divergent, biological one: "Raw water overflows with information. And so the municipality can sell the inert water, they purify it of bugs, data, and matter. With these residues, the biogenetic bank was assembled. This node was from one of the disappeared, a genetic trace that we caught in the river" [El agua cruda trasborda información. Y para que el municipio pueda vender el agua inerte la depuran de bichos, datos y materia. Con esos residuos se armó el banco biogenético. Este nodo era de uno de sus desaparecidos, una traza genética que atrapamos en el río] (107–108). This description alludes to three levels, first, water's ability to connect body fragments (in this case, the disappeared father) and facilitate the rescue, reconstruction, and partial restoration of the past. Therefore, water flow serves as a condition allowing the recovery of absence: what is no longer present and was first forced to disappear and then dissolve; what was compelled to become extinct, that is, a forced extinction.[14] In this sense, water plays a dual role: it acts as a solvent that can dissolve and erase evidence, yet it also serves as a medium and container for preserving proof. Secondly, as a privatized resource, water is exploited here to build a gene bank in the same way that web trackers today gather information about the sites users visit on the internet and analyze their behavior to sell the data to commercial companies, violating users' privacy rights. The extraction "of bugs, data, and matter" purifies the water for its final phase, which is its sale for consumption. Water, therefore, operates as narrated matter that interweaves the politics of disappearance with the privatization of resources and by extension, the collective heritage, encompassing memory and national history. Finally, water is connected to extractive practices within a broader context, as this scenario acts as a parody of another continuity and symbol of the past in the present: Latin America's ongoing dependence on extracting raw materials for sale in global markets since the late nineteenth century. While the hinterlands have been drained and transformed into zones of intense economic activity and extraordinary environmental damage, these conditions have shaped national identities, leading to labels such as "oil nations, banana republics, or plantation societies" and more broadly as underdeveloped or backward nations (Coronil 37). Conversely, the imperial frontier has ingeniously forged new modes of extraction that challenge and redefine traditional territorial boundaries, blurring the lines between country and city and establishing new borders or eliminating them altogether.

If the debate around the Anthropocene, Capitalocene, or Plantationocene, to cite just three terms, aims to identify the origins of the crisis while also assigning responsibility for the conditions that led to its emergence, there is consensus on the various extraction mechanisms that have characterized European, American, Canadian, and Chinese imperialism in all its forms—ranging from mining and deforestation to the exploitation of human and natural resources, the imposition of monoculture worldwide, and the production of waste through planned obsolescence and the overproduction of consumer goods, among others. This procedure is what economist Norman Girvan has defined as "extractive imperialism," a regime of power, ideology, and income distribution generated by resource extraction that has historically revealed both continuities and changes. While the term "Capitalocene," introduced by Jason Moore in 2015 as an alternative to "Anthropocene," emphasizes the role of capitalism in driving anthropogenic environmental changes, the concept of "Plantationocene," also proposed in 2015 by Anna Tsing, highlights the colonial legacy in land use and resource extraction practices in the contemporary era. From the industrial and urban impact on natural and built environments to the agricultural and racialized labor of monoculture, all these terms examine practices that continue to threaten flows of knowledge, biomes and ecosystems, and human and animal bodies across multiple spaces.

13. Forced Extinctions

In Aboaf's novel *El Rey del Agua*, the disappearance of people can also be seen, one might argue, as a form of extraction. The body of the missing father was extracted, "sucked" (*chupado*), as they said in Argentine military jargon, by operating forces that, although they differ in style and scale, reappear today through neoliberal policies where the notion of human and nonhuman expendability is increasingly visible and tangible. Therefore, the politics of extraction illustrate how necropower dynamics persist and evolve through new methods, emphasizing historical continuities in power relations. It is important to clarify that I am interested neither in equating dictatorship with the extractive policies and practices characteristic of contemporary neoliberalism nor in diminishing the severity and scale of the military junta's atrocities, which included the brutal disappearance and torture of thousands of innocent individuals. Instead, my aim is to explore the persistent and evolving manifestations of violence, examining how these forms mutate and re-emerge across different modalities. During the extraction process, the body remnants—tiny particles that mix with others, such as microplastics commonly found on terrestrial and aquatic surfaces—are encapsulated; in doing so, the extraction process reinforces the dividing line that separates nature from culture, the disposable from the usable, the impure from the pure.

In *Une autre science est possible* (*Another Science Is Possible*, 2017), Belgian philosopher Isabelle Stengers addresses the necessity of decelerating scientific progress. She cautions that scientific endeavors, if unmanaged, risk becoming instruments of neoliberal exploitation, with laboratory experiments and research potentially serving the interests of pharmaceutical industries at the expense of the public good. In *El Rey del agua*, Aboaf references this kind of exploitation: "it was enough of a tiny part of the body that traveled through the rivers from Alto Paraná, a sample that they captured in the Delta, to redo the trace. This is how Blanco [the disappeared father] is now here, in a jar in the laboratory, floating in the water" [bastó una minúscula parte del cuerpo que viajó por los ríos desde el Alto Paraná, una parte que captaron en el Delta, para rehacer la traza. Así es como Blanco (el padre desaparecido) ahora está aquí, en un frasco en el laboratorio, flotando en el agua] (115). Just as waste runoff from extractive industries either sinks or floats in geological and aquatic matter, debris becomes inseparable from the colonization processes—geographical, ideological, cultural, and imaginary—of Latin America and, more broadly, of the Global South, creating the conditions for their appropriation. Material remnants flow from peripheral necrospaces to urban centers of distribution and consumption through multiple waterways, binding otherwise disconnected riverscapes and tracing the disposal of debris over countless years.

14. Islands in the Making

In *El ojo y la flor*, the concluding novel of Aboaf's trilogy, the conditions of the hydric landscape undergo a significant transformation. Persistent droughts, coupled with rising higher temperatures, alter the previously aqueous necroscape, rendering it into a semidry quagmire:

> It was no secret that the coastal channel was clogging up and each year it was more difficult to maintain it, but while the water was being exported in Tigre, they knew that the dredging would continue. They had settled for ambiguous news. The oldest clubs were permanently closed, and the sailboats could no longer leave from the moorings. The women and men who walk on the shore or flutter while exercising hide the obvious: the bare mud of water mixed with the garbage is a mirror of the disaster into which nobody wants to look. (123)

> No era ningún secreto que el canal costero se estaba embancando y cada año era más difícil mantenerlo, pero mientras en Tigre se exportaba el agua, sabían que el dragado se mantendría. Se habían conformado con noticias ambiguas. Los clubes más viejos estaban clausurados de manera definitiva, ya no podían salir los veleros desde las amarras. Las mujeres y los hombres que caminan en la orilla o se agitan haciendo ejercicio disimulan lo evidente: el barro desnudo de agua mezclado con la basura es un espejo del desastre en el que nadie quiere mirarse. (123)

The channel is now artificial; on the stairs that connect the dock with the former river, one can see "the old watermarks" [las viejas marcas del agua] (123). According to the narrator, "That channel has been a fiction for a while. And San Isidro has already grown two islands" [Ese canal hace rato que es una ficción. Y a San Isidro ya le crecieron dos islas] (123). Thus, a river that once contained abundant water for industrial extraction and export now stands as a poignant emblem of the colonial legacy in Latin America. It draws attention to the region's persistent pattern of exporting raw materials—ranging from bananas, coffee, corn, cocoa, and oil to minerals, meat, soybeans, and flowers, and now extending to water itself. The spectral river, subjected to increasing anthropogenic depletion through water extraction, exemplifies among many other necrospaces analyzed herein, a seemingly anachronistic archive. This site, referred to as a "wet cemetery" [cementerio húmedo] by the narrator, produces "giant corpses" [cadáveres gigantes] and "dead people" [gente muerta] that emerge, thereby reconfiguring its topography (178). The Tigre River, turned into a "cemetery of water, another abandoned enterprise" [cementerio de agua, otro de los negocios abandonados] by the King of Water, has been "parasitized by whoever wanted to extract its nectar" [parasitado por quien quiso chupar su nectar] (102).

15. Conclusions

While it is imperative to seek innovative and effective methods of reflection and imagination, the aesthetic expressions relevant to the Anthropocene are situated at the intersection where, as Jaime Vindel explains, the "raw experience of reality and the imaginary projections" [la experiencia bruta de la realidad y las proyecciones imaginarias] that link industrial modernity with specific discursive constructs intersect (17). These vectors organize and (re)order extractive itineraries, capturing the dynamics and historical tensions between what exists and what has been extinguished or become extinct, as well as the contrasts between scarcity and abundance.

The anachronistic archives of the Anthropocene encapsulate within their stratifications the remnants of stories that mainstream narratives—those "errors" in chronology, as discussed earlier, characteristic of the framework of modern, teleological rationality—discarded, while simultaneously interrogating and challenging historical paradigms. In this sense, narrative, poetry, and visual art allow for the reorganization of complex interdependencies, entanglements, and dialectical relationships, regardless of their intended aspirations. This is attributable to the fact that the Anthropocene does not manifest solely as a topic, content, or motif, but rather intersects with writing, life, affections, reflexive methodologies, and sensory experiences. Consequently, although necrospace signifies death, it also facilitates the emergence of organic forces as they decompose. These processes serve as catalysts for vitality, revealing a latency that seems absent within the narratives themselves. If geological stratifications of the Anthropocene compose archival layers, then narrative, poetry, and visual art collectively offer avenues to cultivate a situated and ecological ethos, fostering collaborative alliances and generative aesthetic assemblages. In this vein, creative scars, concealed within the hidden crevices of the archive, become vestiges that emerge through synergetic processes, thereby evincing their expressive capacity and generative potential via the materiality itself.

Author's Note

This book chapter is a translation of the article "El necroespacio del Antropoceno: un archivo anacrónico," published in the special dossier "Materialidades del Sur" of the journal *Estudios Filológicos* 72 (December 2023) and coedited by Antonia Viu Bottini and Pedro Eduardo Moscoso Flores. I am grateful to the journal editors for granting me permission to publish the article in English.

Notes

1 While there is undeniable evidence of humans' impact on the planet, a panel of experts in 2024 voted down the proposal to officially declare the start of a new interval of geologic time, one defined by humanity's changes to the planet (Zhong). Whether this transformation should be described as an "event" instead of an "epoch," as currently discussed by scientists, I will continue to use the term "Anthropocene," because of its significance as a marker of the continued exploitation, abuse, and extraction on behalf of human production and consumption.

2 Examples of these effects and phenomena include displacements and deformations, abrupt changes, long interruptions, and discontinuities, all of which reveal a heterogeneous and uneven nature.

3 "Anachronism," Merriam-Webster, https://www.merriam-webster.com/dictionary/anachronism.

4 Another meaning of "anachronism" (*anacronismo*), formulated by the Royal Spanish Academy (Real Academia Española) is "error consisting of confusing eras or placing something outside of its era" [error consistente en confundir épocas o situar algo fuera de su época] (Real Academia Española, *Diccionario de la lengua española,* 23rd ed., versión 23.8, https://dle.rae.es).

5 Besides "Edward Fuchs, Collector and Historian" (first published in 1937), other examples of Benjamin's writings that explore the work of collecting, archiving, and preserving the ephemeral—particularly what resists commodification and market value—through reflections on objects, memory, detritus, and history include *One-Way Street* (1928), "Unpacking My Library" (1931), and the unfinished *The Arcades Project* (1927–1940).

6 It is important to emphasize that this genealogy, far from pretending to be exhaustive, operates as a provocation, an invitation to reflect on its potential articulations. In section 10, for example, I propose, as an interlude, rereading *Pedro Páramo* (1955), by Mexican author Juan Rulfo, as a key reference in the articulation of a counterhegemonic archive.

7 In his "Thesis XVI," Benjamin details how seizing a memory in the present disrupts historicist time and awakens critical consciousness (4:396).

8 Didi-Huberman further discusses German art historian and cultural theorist Aby Warburg's concept of the survival of images (*Nachleben*), and argues that the material used to create images includes "latencies and symptoms, buried and surfaced memories, anachronisms, and critical thresholds, discomfort, and suspensions" [latencias y síntomas, memorias enterradas y memorias surgidas, anacronismos y umbrales críticos] (*Ante el tiempo* 73). Furthermore, in *The Surviving Image: Phantoms of Time and Time of Phantoms* (2017; *L'image survivante: Histoire de l'art et temps des fantômes selon Aby Warburg* [2002]) Didi-Huberman describes how in Warburg's work, "the term Nachleben refers to the survival (the continuity or afterlife and metamorphosis) of images and motifs—as opposed to their renascence after extinction or, conversely, their replacement by innovations in image and motif. . . . Survival entails a complex set of operations in which forgetting, the transformation of sense, involuntary memory, and unexpected rediscovery work in unison. In other words . . . it is a notion that cuts across any chronological scheme; in short, it anachronizes history" (48–49).

9 See also Achille Mbembe, *On the Postcolony.*

10 For a broader understanding of "uneven development," particularly vis-à-vis the unique characteristics of Latin America, see Arturo Escobar's *Encountering Development: The Making and Unmaking of the Third World.*

11 Unless otherwise indicated, all translations are my own.

12 The word *principales* is a term loaded with cultural and historical significance. It refers to the local elite, particularly wealthy landowners, mestizos, or descendants of colonial figures, who wield political, economic, and social influence in Indigenous Andean towns. It can be translated as the landowning class or the powerholders. However, because accurately translating this term into English requires sensitivity to its social, racial, and hierarchical connotations, I decided to keep it in Spanish.

13 *Tertiana* is the Spanish term for tertian fever, a type of intermittent malaria characterized by fever episodes occurring every forty-eight hours. It was common in certain Andean regions and is described in medical and literary texts of the nineteenth and early twentieth centuries (Real Academia Española, *Diccionario de la lengua española,* 23rd ed., versión 23.8, https://dle.rae.es). See also Segrera.

14 I use the term "extinction" following the definition of "termination or end of a thing" (Oxford Languages).

Works Cited

Aboaf, Claudia. *El Rey del Agua.* Alfaguara, 2016.

Aboaf, Claudia. *El ojo y la flor.* Alfaguara, 2019.

Alaimo, Stacy. *Bodily Natures: Science Environment and the Material Self.* Indiana University Press, 2010.

Arguedas, José María. "Agua." In *Relatos completos,* 8-35. Alianza, 1983.

Beck, Ulrich. *Risk Society: Towards a New Modernity.* Sage, 2004.

Benjamin, Walter. *Selected Writings: Volume 3, 1935–1938.* Edited by Howard Eiland and Michael W. Jennings. Harvard University Press, 2002.

Benjamin, Walter. *Selected Writings: Volume 4, 1938–1940.* Edited by Howard Eiland and Michael W. Jennings. Harvard University Press, 2003.

Bjornerud, Marcia. "Wrinkled Time: The Persistence of Past Worlds on Earth." *Emergence Magazine,* October 17, 2024. https://emergencemagazine.org/essay/wrinkled-time/.

Braidotti, Rosi. "Bio-Power and Necro-Politics." https://necrocenenecrolandscaping.wordpress.com/wp-content/uploads/2018/01/braidotti_biopower-and-necropolitics.pdf.

Coronil, Fernando. *The Magical State: Nature, Money, and Modernity in Venezuela.* University of Chicago Press, 1997.

DeLoughrey Elizabeth, and George Handley. *Postcolonial Ecologies: Literatures of the Environment.* Oxford University Press, 2011.

Didi-Huberman, Georges. *Ante el tiempo: Historia del arte y anacronismo de las imágenes.* Adriana Hidalgo, 2011.

Didi-Huberman, Georges. *The Surviving Image: Phantoms of Time and Time of Phantoms, According to Aby Warburg.* Translated by Harvey Mendelsohn. Penn State University Press, 2017.

Escobar, Arturo. *Encountering Development: The Making and Unmaking of the Third World.* Princeton University Press, 1995.

Farrier, David. *Anthropocene Poetics: Deep Time, Sacrifice Zones, and Extinction.* University of Minnesota Press, 2019.

Fornoff, Carolyn, and Gisela Heffes. *Pushing Past the Human in Latin American Cinema.* SUNY Press, 2021.

Girvan, Norman. "Extractive Imperialism in Historical Perspective." In *Extractive Imperialism in the Americas: Capitalism's New Frontier,* edited by James Petras and Henry Veltmeyer, 49-61. *BRILL,* 2014.

Gould, Stephen Jay. *Time's Arrow, Time's Cycle: Myth and Metaphor in the Discovery of Geological Time.* Harvard University Press, 1987.

Gómez, Liliana, and Lisa Blackmore. *Liquid Ecologies in Latin American and Caribbean Art.* Taylor and Francis, 2020.
Huggan, Graham, and Helen Tiffin. *Postcolonial Ecocriticism: Literature, Animals, Environment.* Routledge, 2010.
Iovino, Serenella. "Material Ecocriticism." In *Posthuman Glossary,* edited by Rosi Braidotti and Maria Hlavajova, Bloomsbury Academic, 113–114. 2018.
Iovino, Serenella, and Serpil Oppermann. *Material Ecocriticism.* Indiana University Press, 2014.
Kelly, Jason M. "Anthropocenes: A Fractured Picture." *Rivers of the Anthropocene,* edited by Jason M. Kelly, Philip V. Scarpino, Helen Berry, James Syvitski, and Michel Mey, 1-18. University of California Press, 2018.
Latour, Bruno. "Agency at the Time of the Anthropocene." *New Literary History* 45, no. 1 (2014): 1–18.
Moore, Jason W. *Capitalism in the Web of Life: Ecology and the Accumulation of Capital.* Verso, 2015.
Mbembe, Achille. "Necropolitics." *Public Culture* 15, no. 1 (2003): 11–40.
Mbembe, Achille. *On the Postcolony.* University of California Press, 2001.
Meadows, Lucien Darjeun. "Circling Eloh: A Meditation." *New England Review,* 16 July 2021. https://www.nereview.com/2021/02/11/lucien-darjeun-meadows/.
Oloff, Kerstin. "The 'Monstrous Head' and 'The Mouth of Hell': The Gothic Ecologies of the 'Mexican Miracle.'" In *Ecological Crisis and Cultural Representation in Latin America,* edited by Mark Anderson and Zélia M. Bora, 79–98. Lanham, MD: Lexington, 2016.
Pratt, Mary Louise. *Planetary Longings.* Duke University Press, 2022.
Quiroga, Horacio. "The Log-Fishermen." In *The Latin American Ecocultural Reader,* translated by J. David Danielson, edited by Jennifer French and Gisela Heffes, 138-142. Northwestern University Press, 2021.
Rivera Garza, Cristina. *Los muertos indóciles: Necroescrituras y desapropiación.* Penguin Random House, 2019.
Rulfo, Juan. *Pedro Páramo.* Fondo de Cultura Económica, 1955.
Segrera, Claudio. *Enfermedades en los Andes: Historia médica del Perú rural.* Universidad Nacional Mayor de San Marcos, 2002.
Smith, Neil. *Uneven Development: Nature, Capital, and the Production of Space.* University of Georgia Press, 2008.
Stengers, Isabelle. *Une autre science est possible.* La Découverte, 2017.
Tsing, Anna Lowenhaupt. *The Mushroom at the End of the World: On the Possibility of Life in Capitalist Ruins.* Princeton University Press. 2015.
Vindel, Jaime. *Estética fósil: Imaginarios de la energía y crisis ecosocial.* Arcàdia/Macba, 2020.
Wizisla, Erdmut. Preface to *Walter Benjamin's Archive: Images Texts Signs,* edited by Ursula Marx, Gudrun Schwarz, Michael Schwarz, and Erdmut Wizisla, 6-9. Verso, 2015.
Zhong, Raymond. "Are We in the 'Anthropocene,' the Human Age? Nope, Scientists Say." *New York Times,* March 6, 2024. https://www.nytimes.com/2024/03/05/climate/anthropocene-epoch-vote-rejected.html.

8

The Ecological Novel

Unearthing a Latin American Land Archive

Carlos Fonseca

The concluding pages of Gabriel García Márquez's *Cien años de soledad,* marked as they are by the prophetic tone of Melquíades's manuscripts, combine a sense of return with a fatal ending. As Aureliano Babilonia and Amaranta Úrsula, the last members of the Buendía family, begin to realize their tragic destiny, nature encircles them like it did in the early days of Macondo.

> The silver shop, Melquíades' room, the primitive and silent realm of Santa Sofía de la Piedad remained in the depths of a domestic jungle that no one would have had the courage to penetrate. Surrounded by the voracity of nature, Aureliano and Amaranta Úrsula continued cultivating the oregano and the begonias and defended their world with demarcations of quicklime, building the last trenches in the age-old war between man and ant. (*One Hundred Years* 377)

In those last pages, as the story reaches its end, nature overtakes culture, reclaiming for itself the space from which it had been expelled under the pretense of history, civilization, and progress. However, as we soon witness, this return of nature is not that of the peaceful paradisiacal garden but rather of a catastrophic nature that under the form of a biblical cyclone that knows no boundaries, reduces Macondo to a heap of rubble. Today, when climate change seems to reduce history to a litany of natural catastrophes, and when, as Dipesh Chakrabarty has explained, the boundary between natural history and human history appears to have collapsed, this ending seems prophetic. After many years of being profoundly urban, many contemporary Latin American novels are characterized by a return to nature or at least by a profound ecological concern. However, as in García Márquez's novel, this nature is never Edenic but is rather traversed by a precarious historicity that more often than not operates under the sign of catastrophe.

Many contemporary novelists have inherited this ruinous landscape where nature entangles itself around the vestiges of history like morning glory. In the nineteenth century, Argentine President Domingo F. Sarmiento had trusted the vector of history, the supposed linear development of civilization, to subsume the barbarism of nature, as posited in his famous 1845 work *Facundo.* If in those closing pages of *Cien años de soledad* Sarmiento's dream of urban progress had suddenly turned into a nightmare, with the purported barbarism of nature quickly overtaking civilization, many contemporary writers find that it is precisely this catastrophic dialectic, this necessity to bear witness to what Elizabeth Povinelli has called the "ancestral catastrophe," that must be explored if the world is to survive beyond the many apocalyptic ends being projected upon it. They understand that to write about and from Latin America is to bear the weight of this complex dialectic on their shoulders. The imaginary of Sarmiento's pampas is in their minds as is the long-standing history through which Latin America has been defined by its double bind to nature. Contemporary novelists rewrite the complex dialectic through which Latin America has been defined since colonial times, both by a desire for nature and a fear of it. This double knot, where dream and nightmare coincide, is everywhere in literature. It is the desire for gold and spices of Columbus turning into the fear of cannibals; it is the perverse desire for the marvelous that confronted the colonial naturalists; it is the catastrophic sublimity that fascinated Alexander von Humboldt and the seductive double edge of captivity narratives. It is, at the end of the day, the complex dialectic between dream and nightmare, desire and fear, that lies behind the closing phrase of José Eustasio Rivera's 1924 novel *La vorágine:* "The jungle has swallowed them!" (*The Vortex* 379). Today, almost a century after Rivera wrote that fateful sentence, its resonance bears witness to the romantic seductions of present-day ecotourism as well as to a long history of ecological exploitation and extractivism in the Global South. For it has been this complex tapestry of desires and fears that has posited Latin America and more broadly the postcolonial world as a space of extraction and terror, as a space of *materia prima* and underdevelopment. Raymond Williams has famously stated, "The idea of nature contains, though often unnoticed, an extraordinary amount of human history" (*Ideas of Nature* 67). Waking up to this realization, the contemporary novelist tries to read the historical traces that have been covered under the mask of nature.

Paradoxically, nature has become our most precious commodity and our curse. Confronted with this double bind, which acts both as a fantasy and a painful reality, the contemporary Latin American novelist realizes the political complexity of their task. They must disarticulate this fantasy from within, paying attention to how it has legitimized how others see the region and how it envisions itself. The task is therefore not simple, as it is never simple to deal with a fantasy, for as Michael Taussig has remarked, doing so requires penetrating "the hallucinatory veil of the heart of darkness without either succumbing to its hallucinatory quality or losing that quality" (496). Staying clear of both romantic exoticism and rationalist simplification, the gaze of the contemporary novelist must navigate against the current of this ecological fantasy, always conscious of the painful realities at stake in this reimagining of nature against the grain.

The writers have, however, at their disposal the best tool: stories. If, as Taussig himself explains in his essay "Culture of Terror—Space of Death," these colonial fantasies were established through narration and storytelling, it might then be that one of the strongest weapons for getting out of them might be stories and narrative. With the logic of the *pharmakon* that often characterizes art and literature whereby the literary remedy "cites, re-cites, and makes legible that which *in the same word* signifies, in another spot and on a different level of the stage, *poison*" (Derrida 100), the task of the ecological novel is that of traversing the perverse dreams of the ecological fantasy that has long sustained extractivist modernity and disarticulating its underlying narratives. In what follows, sketching a tradition of Latin American ecological literature that is as old as it is essential, and pairing it with more recent Latin American novels such as Cristina Rivera Garza's *Autobiografía del algodón,* Vanessa Londoño's *El asedio animal,* Juan Cárdenas's *Peregrino transparente,* and Rita Indiana's *La mucama de Omicunlé,* I would like to explore what this task of rewriting and rereading nature could mean today. The task at hand is not so much to analyze these contemporary novels in detail but rather to see how, in their ecological proposals, they allow us to reread, against the grain, canonical Latin American novels such as García Márquez's *Cien años de soledad,* Alejo Carpentier's *Los pasos perdidos,* and Rivera's *La vorágine.* Adopting today's ecological stance, these canonical texts can finally be read as land archives that bear a critical potential. Such a retrospective reading opens the space for a reconsideration of what ecology has meant and means within Latin American literature. Following closely Rivera Garza's concept of *escrituras geológicas,* geological writings, I will try to think through what the poetics of such an ecological novel might look like, keeping in mind what Rivera Garza asserts.

> Geology is not only a field of knowledge but, more generally, a technology of matter, a racialized and colonialist praxis that goes hand in hand with the processes of extraction and dispossession that have dismantled

entire regions of the planet, expelling native populations and enslaving Black or Native bodies whom, since then, an indifferent geo-logic has categorized as inert matter, that is, nonhuman.

> La geología no solo es un campo de saber, sino, más generalmente, una tecnología de la materia, una praxis racializada y colonialista que va mano a mano con los procesos de extracción y desposesión que han desmantelado regiones enteras del planeta, expulsando a poblaciones nativas y esclavizando a cuerpos negros o nativos a quienes, desde entonces, una geo-lógica indiferente categorizó como materia inerte, es decir, no humana. ("Escrituras" 12)[1]

An ecological novel worthy of its name would then be able to simultaneously address the ecological imaginary and the material conditions of exploitation that such a fantasy made possible. It would be in dialogue with the *novelas de la tierra* and with the colonial chronicles; it would respond to Sarmiento's novel *Facundo* and Werner Herzog's 1982 film *Fitzcarraldo;* it would exhume the specters of violence buried within a landscape that refuses to forget. For, as the concluding pages of *Cien años de soledad* suggest, below the apparent surface of today's precarious landscapes, buried by the progressive passage of time, lies a ruinous history that requires retelling.

Reading Strata: Land Archives of the Contemporary Novel

The first thing that the contemporary Latin American novelist realizes is that there is nothing more unnatural than nature. Nature is neither the Edenic garden outside of history nor the origin that grants authenticity, but rather a space traversed by writings, erasures, and rewritings. As Jens Andermann, Gabriel Giorgi, and Victoria Saramago note in the introduction to their handbook on Latin American environmental aesthetics, "The many pasts and presents of coloniality are, we might say, written on the environment. Critical interventions from environmental aesthetics operate with these traces that we can read today not only in racial and cultural terms but also in the deep time of geological scales" (12). Once they realize this, contemporary novelists feel at home, for they understand that nature is nothing but an archive of human and nonhuman stories. Nowhere is this more evident than in the story of the genesis of what could perhaps serve as an archetypical ecological novel: Carpentier's *Los pasos perdidos* (*The Lost Steps*). The Cuban writer mentioned in interviews that the novel sprang from a series of journeys he made into the Venezuelan wilderness in the 1940s, inspired in part by his childhood reading of Emilio Salgari's jungle books. And it is precisely books that Carpentier found in his journey to the heart of the jungle. In a 1977 interview on Spanish television, Carpentier recalled his visit to the famous waterfall El Salto del Ángel and described the natural landscape but also the bookish characters he found along the way: "On that trip, I met a number of very interesting people: a Greek diamond digger who traveled with the *Odyssey* in his pocket and a book by Xenophon, *The Retreat of the Ten Thousand,* and he read us fragments in Greek while traveling up the Orinoco" [Realizando ese viaje me tropecé con una cantidad de gente interesantísima: un griego buscador de diamantes que viajaba con la *Odisea* en un bolsillo y un libro de Jenofonte, *La retirada de los diez mil,* y nos leía fragmentos en griego remontando el Orinoco] ("A fondo"). At the very heart of the jungle, Carpentier encountered not the outside of literature but rather its very core. In a later essay, "Visión de América," collected alongside other critical writings in the book *Los pasos recobrados,* Carpentier would explain that it was the unexpected encounter with a sort of natural writing—the inscription of a V in a tree trunk—that prompted the epiphany that led him to write the novel.

> I entered through the Caño de la Guacharaca, where I saw the three V-shaped incisions that I paint in my novel. And I remember that one afternoon, at the confluence of the Orinoco and the Vichada, on a luminous, extraordinary afternoon, I had something like an illumination: the novel *Los pasos perdidos* was born in a few seconds, completely constructed, structured, done; all I had to do was return to Caracas and write it.
>
> Entré por el Caño de la Guacharaca, donde vi las tres incisiones en forma de V que yo pinto en mi novela. Y recuerdo que una tarde, en la confluencia del Orinoco y del Vichada, en una tarde luminosa, extraordinaria, tuve algo así como una iluminación: la novela *Los pasos perdidos* nació en pocos segundos, completamente construida, estructurada, hecha; no tenía más que volver a Caracas y escribirla. (*Ensayos* 62)

A similar intuition traverses *Los pasos perdidos.* As the protagonist, an anonymous ethnomusicologist, attempts to leave history behind, writing nonetheless appears ubiquitously. Nature shows itself to be the ultimate archive, a space of inscription, impression, and erasure that is everywhere marked by the violence of history.[2] Carpentier might have wished to find the uneventful nature of the pre-Adamic Genesis, but everywhere he went he was reminded of the inescapability of history. It is not surprising, then, that in Roberto González Echevarría's classic book on the subject, *Mito y archivo,* he calls *Los pasos per-*

didos the archetypical archival novel, for at its core there lies a fundamental intuition that traverses contemporary attempts to write ecologically. It is that nature is an archive and the task of the contemporary writer is to learn, like Carpentier in front of the inscribed tree, to read nature's enigmatic hieroglyphs.

This insight is not necessarily new. It is something that became evident to the geohistorians of the eighteenth century—people like Cuvier, Buffon, and Desmarest—when they finally left behind the theodicean myth of the deluge and started speaking of nature's documents and archives. The discovery of nature's historicity, accompanied by the geological establishment of the figure of the fossil and strata, becomes today a matter of literary relevance. Like Carpentier in front of the tree trunk, the contemporary novelist realizes that the concept of writing extends beyond the realm of humanity. In Rivera Garza's concept of *escritura geológica,* geological writing, nature is imagined as a historical archive: "A geological writing is thus proposed, on principle, as a desedimentative operation. Geology, on the other hand, constantly reminds us that we are time" [Una escritura geológica se propone así, por principio de cuentas, como una operación desedimentativa. La geología, por otra parte, nos recuerda constantemente que somos tiempo] (12). To write ecologically then means, first and foremost, to act a bit like an archaeologist or a geologist, unearthing the historical layers of meaning that have been successively projected onto the apparent atemporality of the natural landscape. Yusoff has observed that "stratigraphy went well beyond the natural sciences and extractive geosciences and saw a flourishing in ideas of social theories through the nomenclature and metaphor of the formation as the our-come and process of social, political, and economic forces" (242). Unearthing such layers and disarticulating the fantasies that are at stake in each of them becomes the political task of the ecological writer. Rivera Garza finds that landscapes, as sediments, bear an archive of stories waiting to be told.

> The present is but the most recent sediment and, therefore, the most superficial—the tip of the iceberg, as Hemingway would say—which announces, although it does not allow us to see fully, the multiple layers that, superimposed one on top of the other, constitute a past that is never lost but is preserved in rocks, landscapes, glaciers, and various ecosystems. The Earth is thus our first great geological archive: the repository of initiatory experiences and the last ones as well.

> El presente no es sino el sedimento más reciente y, por lo mismo, el más superficial—la punta del iceberg, diría Hemingway—que anuncia, aunque no permite ver a cabalidad, las múltiples capas que, sobrepues-

> tas una sobre otra, constituyen un pasado que nunca se pierde, sino que se conserva en rocas, paisajes, glaciares y ecosistemas varios. La Tierra es, así, nuestro primer gran archivo geológico: el repositorio de las experiencias iniciáticas, y las últimas también. (12–13)

Through the concept of *escrituras geológicas*, Rivera Garza indirectly proposes a poetics of the geological novel in which the seeming superficiality of the landscape is replaced by the historical depth of what could perhaps be called the land archive. Against the framed notion of the landscape, which for Raymond Williams "always implies separation and distance" (*The Country* 120), and whereby landscape is imagined as a canvas for projection and admiration, the idea of a land archive suggests an unframing of nature and an archaeological exploration of its temporal constitution. Jens Andermann has also noted that "landscape, in short, represents a key ideological apparatus of capitalism and colonialism that naturalizes what are in fact violent and uneven social and political relations" (9). If nature is a sedimented layering of historical strata, the act of writing becomes what Rivera Garza calls "desedimentación," the progressive critical unearthing of those layers and the disarticulation of their exploitative fantasies. Attentive to what Carpentier once called the novel's "chthonic context" (*contexto ctónico*), Rivera Garza's approach immerses the landscape in its long-standing political context. Like eighteenth-century geohistorians, Rivera Garza suggests reading nature less from the punctual perspective of the forgetful present than from a perspective that remains faithful to what James Hutton, the pioneer of modern geology, first termed "deep time."

Replacing landscapes with land archives, the ecological novel explores the *longue durée* histories of ecological violence buried deep within the shiny superficiality of postcard nature. In doing so, these novels escape the presentism of most ahistorical depictions of the environment in which the lack of a historical approach allows for its uninterrupted exploitation at the hands of a capitalist modernity that only cares about today's profits. Instead, seeing nature as a land archive opens a view of the ways ecological violence has been exerted, in the sensationalist fashion so beloved by the media such as oils spills in the Pacific and wildfires in California but also in a slower manner that nonetheless has horrible consequences for large areas of the planet. Treating canonical texts geologically, that is to say, reading them as stratified archives composed of sedimented meanings and fantasies, unearths the true temporality of environmental violence. That "slow violence," as Rob Nixon has called it, might remain ungraspable and invisible if, accustomed as we are to human time, we refuse to see things from the perspective and scale dictated by nature itself. Only then will we understand what the first geohistorians of the eighteenth century might have felt when they broke a rock in the middle of the desert and saw, in its transversal strata, that nature was nothing but the concretization of time.

On the Temporal Flows of the Ecological Novel

The rivers that flow through recent ecological novels are not that different from the river of Heraclitus; in them as well, nature is profoundly traversed by the flow of time. More often than not, however, these temporal flows are read as a journey into the past. Nature, conceived as an original source of authenticity and autochthony, is posited as the lost past from which civilization has expelled humans and to which we must return. The task remains for the contemporary ecological novel to show that while this journey to the seed sets out to explore the past, it also opens up the possibility of imagining new futures. *Los pasos perdidos* is again illuminating in this sense, for it sketches the limits of these primitivist narratives of nostalgic return. Carpentier said that when he decided to row up the Orinoco in 1944, his goal was to reach the outside of time, that original source of culture that stood at the very beginning of everything. This is the search he undertakes in his unfinished chronicles of *El libro de la Gran Sabana,* and this is the desire that pushes the protagonist of *Los pasos perdidos* to leave the city behind and to venture into the jungle. Indeed, at first, the novel seems to be an ode to the authenticity of that forgotten world outside time: "For two days we had been crawling along the skeleton of the planet, forgetting History and even the obscure migrations of the unrecorded ages. . . . What lay before our eyes was the world that existed before man" (*The Lost Steps* 185–186). The ethnomusicologist protagonist's quest to retrieve a series of musical instruments is embedded in his search for a past origin that would contrast the speedy advancement of time within modernity. In many ways, the novel tells the story of that utopia outside of the constraints and norms of modern society, which Carpentier inherited from the project of the avant-garde. Santa Mónica de los Venados is a Valle del Tiempo Detenido (Valley of Stopped Time), the Garden of Eden. *Los pasos perdidos* would be a simple primitivist utopia if it were not for Carpentier sketching within its pages the dissolution and critique of such a fantasy.

Forced to return to the city, the protagonist's journey of return is not as successful as the outbound one. Nature, no longer behaving passively, blocks his path into Eden. The V-marked tree that had once guided his way is no longer visible, the mark having been erased by the rains. Nature begins to show its true face as historical agent: "The hut where Rosario and I knew each other for the first time had been literally burst apart by the force of the plants that had grown into it, pushing up the roof, cracking the walls, turning to dead leaves, rot, what once had been the materials of which a home was made" (267–268). Like the catastrophic nature García Márquez presents in the concluding pages of *Cien años de soledad,* the force of the natural begins to puncture here the seemingly ahistorical fabric of the landscape, inscribing history back in the picture. *Los pasos perdidos* shows and disarticulates the limits and impasses of the primitivist dream of return while opening the temporal horizon of the ecological novel. Always traversed by history, the ecological novel must go beyond a nostalgic desire for origins; it must arrive, like the protagonist of Carpentier's novel and the last members of the Buendía family, at a revelation that the only path to the forbidden Garden of Eden is found by facing the future.

Future-facing, contemporary ecological novelists have learned the lesson. They know well that the gate to the Garden of Eden is closed and that the past is accessed by way of the future. Rivera Garza has explained how these ecological writings critically project what was into what will be.

> Far from being a nostalgic gesture, dreaming of a past in which everything was better, these authors test and remove, cut and intermingle, doing, in short, everything possible to open that crack in the present through which will burst, with all its future critical power, the past that lives under our feet or flies in the atmosphere along with the air we breathe.
>
> Lejos de ser un gesto nostálgico, que sueña con un pasado en que todo fue mejor, estos autores testerean y remueven, cortan y entremezclan, haciendo, en fin, todo lo posible para abrir esa grieta en el presente por donde irrumpirá, con toda su futura potencia crítica, el pasado que pervive bajo nuestros pies o vuela en la atmósfera junto con el aire que respiramos. (15)

In the archival process of writings and rewritings, readings and rereadings that characterize these *escrituras geológicas,* nature no longer appears as the Latin American horizon of ontological meaning but rather as a critical space of becomings.[3] No longer the space of origins, authenticity, and being, nature becomes the contested space where political narratives are articulated and disarticulated. In this sense, it remains a utopian space less for its position outside of history than precisely because of its redemptive position within cultural history. Carpentier's texts are positioned at the transition point between two ways of thinking anthropologically; *Los pasos perdidos* marks the novelistic shift from a conception of culture that finds in nature a site for identarian roots to a conception of nature as a transversal space marked by routes. This transition from roots to routes, in James Clifford's assessment, marks the development of anthropology in the twentieth century; the transition helps novelists resituate nature within the temporal field. Nature is no longer anchored in the rooted past; it is a site traversed by routes that point to the future. If the classic *novelas de la tierra* worked under the old ontological paradigm of origins, authenticity, and autochthonous roots, contemporary rewritings like those provided by Edmundo Paz Soldán and Gabriela Cabezón Cámara reimagine these landscapes as sites of beginnings, becomings, and routes. To narrate nature means traversing it and finding out what lies on the other side of the fantasy.

Traversing the Fantasy: Journeys across the Natural

The ecological novel is tied from its very beginnings to the image of the journey. From early influences such as Humboldt's *Personal Narrative of Travels to the Equinoctial Regions of the New Continent* and Lucio Mansilla's *Una excursión a los indios ranqueles,* through the canonical *novelas de la tierra* such as Rivera's *La vorágine* and Rómulo Gallegos's *Doña Bárbara,* all the way to more recent *escrituras geológicas* such as Cárdenas's *Peregrino transparente* and Cabezón Cámara's *Las aventuras de la China Iron,* the tradition of the ecological novel is one of journeys, expeditions, voyages, adventures, crossings, and encounters.[4] To say it with a Spanish word that perhaps bears more resonance with what I have been trying to think through: its tradition is full of *travesías,* of crossings. To think of the ecological novel as a genre of *travesías* means to remain faithful to what is disclosed by the term itself, that is, the idea of crossing or traversing nature and finding something on the other side. What is it, then, that these novels find when, having cut across the space of the jungle, the plain, or the desert, they finally reach the other side? What is it, what vision, intuition, or illumination do they bring back to the city once their natural journeys conclude?

As *Los pasos perdidos* shows, and as we find in the books of Humboldt, Darwin, and Fernández de Oviedo, these questions are not easy to answer precisely because they mix the history of knowledge with the history of fantasy. Whether it is science, anthropology, or ethnography, these journeys are often performed in the name of objective knowledge but end up producing something else, a dreamlike shadow that extends over knowledge as its menacing double. Knowledge and fantasy entwine around each other with the same baroque promiscuity that Carpentier assigns to Caribbean nature in his famous prologue to the novel *El reino de este mundo.* In that work Carpentier casts the question of representing nature in terms of a complex poetics of entanglement that overwhelms Western perception.

> But observe that when André Masson tried to draw the jungle of Martinique, with its incredible intertwining of plants and its obscene promiscuity of certain fruit, the marvelous truth of the matter devoured the painter, leaving him just short of impotent when faced with blank paper. It had to be an American painter—the Cuban, Wifredo Lam—who taught us the magic of tropical vegetation, the unbridled creativity of our natural forms with all their metamorphoses and symbioses on monumental canvases in an expressive mode that is unique in contemporary art. (*Kingdom of This World* xv)

This passage, which refers to the famous disembarkment of the *Capitaine Paul Lemerle* in Martinique and the journey of its avant-garde passengers to the forest of Absalon, accurately inscribes the problem of nature and its representation within the world of fantasy as well as in positive knowledge. Carpentier contends that what distinguished Lam from Masson in their attempts to capture the baroque excess of nature was not so much the Cuban artist's capacity to see nature as it truly was but rather his talent for hinting at its fantastic core. It is in this prologue that Carpentier proposes his theory of the *real maravilloso americano,* the marvelous real of the Americas. Indeed, what seems to distinguish Lam's canvas after his return to Cuba is his capacity to portray nature as a space both of realities and fantasies, of dreams and nightmares. When looking Lam's paintings *The Jungle* or *Dark Malembo, God of the Crossroads* (both in 1943), what becomes clear is that within them, tropical nature—too overwhelming for positivist description—begins to lead the rational mind astray into a descriptive madness where desire and fear coincide. Lam's talent lay in his capacity to paint the specters of those colonial fantasies without giving himself over to them, in a way penetrating the hallucinatory veil depicted by Taussig. Instead, as Carpentier suggests, Lam was able to traverse the colonial fantasy enveloping nature and find, on its other side, the possibility of a postcolonial awakening.

This face-off with the colonial fantasy at the core of the natural journey is everywhere within the tradition of ecological writing in or about Latin America. It is Darwin's fascination with the landscape and people of Tierra del Fuego; it is the catastrophist language of Humboldt's most poetic pages; it is the orientalist pampa of Sarmiento; it is the devouring jungle of Rivera. In each case, nature is enveloped in a veil of fantasy within which desire and fear coincide and confuse themselves. As Lam's example showcases and Taussig's "heart of darkness" quote underlines, it is important not to shy away from such a fantasy but rather to face and traverse it. The task of the ecological novel is to understand how this fantasy lies at the heart of the extractivist dream-turned-nightmare that even today gets projected onto the region. The task is to place the reader in front of the dialectic of desire and fear that has long sustained the ecological dreams of extractivist modernity and to propose, as Walter Benjamin has suggested, an "awakening."

From Pola Oloixarac's *Las constelaciones oscuras* to Rita Indiana's *La mucama de Omicunlé,* many of Latin America's ecological novels stage such an awakening precisely against the background of a renewed extractivism. As a consequence, their journeys face the past and primordially the future. Finding in the marvelous real an already exhausted aesthetic, their novels turn to technofuturism and science fiction more generally as genres that can help us figure out what traversing our ecological fantasies might mean today. It is perhaps in Paz Soldán's novel *La mirada de las plantas* that the fantastical core of the ecological becomes clearer. In a novel that always remains close to Rivera's *La vorágine,* Paz Soldán focuses on new technologies that resonate with the history of extractivism in the Amazon. The novel follows the story of a laboratory set in the frontier between Bolivia and Brazil where a foreign company is giving people a hallucinogen called *alita del cielo* that allows agents of the company to take control over the patients' subconscious. Weaving complex reflections on data extraction, fake realities, and ecological violence, the novel turns the classical trope of the ecological journey inside out and the *travesía* becomes a mental one, a hallucinogenic trip to the "end of nature," where nature shows its most unnatural side, its fantastic core. Nature and technology intertwine in the novel to show what traversing the ecological fantasy might mean today and how it might foreshadow an awakening and a new beginning.

Conclusion: New Beginnings

Against the tidal wave of apocalyptic endings that permeate the present-day historical imagination, the contemporary ecological novel has been tasked with imagining new beginnings. But how would these new beginnings differ from the genre's classical obsession with Edenic notions of origins, autochthony, and authenticity? What would it mean to speak nowadays of a postautochthonous ecological novel? What would nature even be within a society where the natural and the virtual are, as Paz Soldán suggests, so inextricably interlaced? These are some of the questions that contemporary ecological novels face when they narrate the violent histories of Latin American landscapes. Moreover, staged as they are against a background of ubiquitous catastrophe, their imagined beginnings are always performed after disaster.

This feeling of writing after the end—the end of history, the end of nature, the end of literature—is perhaps the decisive trait of contemporary literary production. To return to the initial image from García Márquez's *Cien años de soledad,* contemporary ecological writers are writing after the end of Macondo, from within the ruins left behind by the biblical cyclone that reduces the town to a heap of rubble. They are writing from within that space where the ruins of the lettered dream of progress become confused with the remnants of Melquíades's scrolls, lost as they are "among the prehistoric plants and steaming puddles and luminous insects that had removed all trace of man's passage on earth from the room" (*One Hundred Years* 381). No longer held together by either the baroque aesthetics of the marvelous real or by the teleology of a continental grand narrative, their stories are proper writings of disaster, like the ones Maurice Blanchot once imagined. The writers' capacities lay in being able to reconfigure, out of the remnants left behind by catastrophe, a new beginning. One must not forget that Macondo itself was founded by the remnants of catastrophe, the product of "a whirling leaf storm [that] had been stirred up, formed out of the human and material dregs of other towns, the chaff of a civil war that seemed ever more remote and unlikely" (*One Hundred Years* 1). This leaf storm, which both founds Macondo and marks its calamitous final destruction, blurs the distinction between beginnings and endings by imagining a town that was always already, from the very get-go, a product of the aftermath of catastrophe.

> In less than a year it sowed over the town the rubble of many catastrophes that had come before it, scattering its mixed cargo of rubbish in the streets. And all of a sudden that rubbish, in time to the mad and unpredictable rhythm of the storm, was being sorted out, individualized, until what had been a narrow street with a river at one end and a corral for the dead at the other was changed into a different and more complex town, created out of the rubbish of other towns. (1)

Writing after the end of Macondo, these ecological writers are well aware that they stand amid a ruinous landscape. They are conscious that the myths of autochthony, authenticity, and origins for which Macondo often stood as a sort of symbol have been long disarticulated by the winds of a progressive modernity that has conflated the end of history with the end of nature. Against such an apocalyptic imaginary, as in the ending of *Cien años de soledad,* their texts sketch the return of nature *as* history. Within their postcatastrophe texts, nature returns as a force that opens a historical horizon where history seems to have exhausted its exits. Mindful of the Anthropocene and the role of humans as agents of geological destruction, the writers nonetheless do not overstress the centrality of the human. Instead, they seek to awaken readers to a complementary intuition, that nature is also historical.

The beginnings that these ecological novels project are to be thought of in terms of awakenings, rewritings, survivals, and returns. Christina Sharpe's work on the postcolonial wake as mourning and as a coming into consciousness might give a closing image or figure through which to think of the ecological poetics that these writers are sketching, each in their own way. Today's ecological novels seem to be written in the wake of nature, mourning its disappearance but awakening to its return. In Latin America, where the projection of utopias has often led to terrible dystopias, it is encouraging to think that, for once, it might be time to turn a dystopian horizon into a utopia to come.

Notes

1 Unless otherwise indicated, all translations are my own.

2 Charlotte Rogers has insightfully remarked in this sense, "Alejo Carpentier's *Los pasos perdidos* . . . is emblematic of this transition: his protagonist at first extols the forest as a pristine and archaic wilderness, but by the end of the novel he recognizes that the intrusion of extractive industries is rapidly transforming the landscape and its peoples. While the tropics occasionally push back against the intruders, the unruly landscape has largely been tamed and no longer offers a new source of artistic inspiration. Instead, these writers chronicle the increase in human settlement, building of roads, clearing of land, and harvesting of the region's resources in the second half of the twentieth century" (44).

3 Gabriel Giorgi observes in his chapter on "Strata": "The novel—in its interface with the extractive frontiers that functioned as violent laboratories of the sensible in Latin America—then, was always haunted by non-human perspectives and scales. . . . Mineral, animal, vegetal agencies and forces bring their own temporalities to narrative" (384).

4 This should not be understood in the sense that the ecological novel is always thought of from the perspective of the colonial gaze. To the contrary, what is at stake here is understanding the ecological journey in connection to a poetics of relation not so distinct from that sketched out by Édouard Glissant. The ecological *travesía,* crossing or traversing, is always an attempt to find an identity via the other in a proper postcolonial fashion. The possibility of a pure, isolated identitarian discourse is precisely what the ecological novel challenges.

Works Cited

Andermann, Jens. *Entranced Earth: Art, Extractivism, and the End of Landscape.* Northwestern University Press, 2022.

Andermann, Jens, Gabriel Giorgi, and Victoria Saramago. Introduction to *Handbook of Latin American Environmental Aesthetics,* edited by Andermann, Giorgi, and Saramago, 1–23. De Gruyter, 2023.

Benjamin, Walter. "On the Concept of History." *Selected Writings: Volume 4, 1938–1940,* edited by Howard Eiland and Michael W. Jennings, translated by Howard Eiland, 401–425. Harvard University Press, 1996.

Blanchot, Maurice. *The Writing of Disaster.* University of Nebraska Press, 2015.

Cabezón Cámara, Gabriela. *Las aventuras de la China Iron.* Random House, 2017.

Cárdenas, Juan. *Peregrino transparente.* Periférica, 2023.

Carpentier, Alejo. "A fondo: Entrevista con Alejo Carpentier." Interview by Joaquín Soler Serrano. Recording. Radio y Televisión Española, 1977. Posted to YouTube, https://www.youtube.com/watch?v=inF8qRk4RDU.

Carpentier, Alejo. *Los pasos recobrados: Ensayos de teoría y crítica literaria.* Biblioteca Ayacucho, 2003.

Carpentier, Alejo. *The Lost Steps.* Translated by Harriet de Onís. Farrar, Straus and Giroux, 1989.

Carpentier, Alejo. "On the Marvelous Real in America." In *Magical Realism: Theory, History, Community,* edited by Lois Parkinson Zamora and Wendy B. Faris, 75–88. Duke University Press, 1995.

Carpentier, Alejo. *The Kingdom of This World.* Translated by Pablo Medina. Farrar, Straus and Giroux, 2017.

Chakrabarty, Dipesh. *The Climate of History in a Planetary Age.* University of Chicago Press, 2021.

Clifford, James. *Routes: Travel and Translation in the Late Twentieth Century.* Harvard University Press, 1997.

Darwin, Charles R. *Journal of Researches into the Natural History and Geology of the Various Countries Visited by H.M.S. Beagle.* John Murray, 1890.

David, Catherine. *Wifredo Lam: The EY Exhibit.* Tate, 2016.

Derrida, Jacques. *Dissemination.* Translated by Barbara Johnson. Continuum, 2008.

García Márquez, Gabriel. *Leaf Storm and Other Stories.* Translated by Gregory Rabassa. Picador, 1972.

García Márquez, Gabriel. *One Hundred Years of Solitude.* Translated by Gregory Rabassa. Bard, 1971.

Giorgi, Gabriel. "Strata." In *Handbook of Latin American Environmental Aesthetics,* edited by Andermann, Giorgi, and Saramago, 381–394. De Gruyter, 2023.

Giorgi, Gabriel. *La mirada de las plantas.* Almadía, 2022.

Glissant, Édouard. *Poetics of Relation.* University of Michigan Press, 1997.

González Echevarría, Roberto. *Mito y archivo: Una teoría de la narrativa latinoamericana.* Fondo de Cultura Económica, 2000.

Indiana, Rita. *La mucama de Omicunlé.* Periférica, 2015.

Humboldt, Alexander Von. *Personal Narrative of Travels to the Equinoctial Regions of the New Continent: During the Years 1799–1804.* Translated by Helen Maria Williams. Cambridge University Press, 2011.

Hutton, James. *Theory of the Earth: With Proofs and Illustrations, in Four Parts.* Geological Society of London, 1977.

Londoño, Vanessa. *El asedio animal.* Almadía, 2021.

Mansilla, Lucio V. *Una excursión a los indios ranqueles.* Imprenta, litografía y fundición de tipos Belgrano 126, 1870.

Nixon, Rob. *Slow Violence and the Environmentalism of the Poor.* Harvard University Press, 2011.
Oloixarac, Pola. *Las constelaciones oscuras.* Literatura Random House, 2016.
Paz Soldán, Edmundo. *La mirada de las plantas.* Almadía, 2022.
Povinelli, Elizabeth. *Between Gaia and Ground: Four Axioms of Existence and the Ancestral Catastrophe of Late Liberalism.* Duke University Press, 2021.
Rivera, José Eustasio. *The Vortex.* Translated by E. K. James. Putnam, 1935.
Rivera Garza, Cristina. *Autobiografía del algodón.* Random House, 2020.
Rivera Garza, Cristina. *Escrituras geológicas.* Iberoamericana, 2022.
Rogers, Charlotte. *Mourning El Dorado: Literature and Extractivism in the Contemporary American Tropics.* University of Virginia Press, 2019.
Sarmiento, Domingo Faustino. *Facundo.* Imprenta del Progreso, 1845.
Sharpe, Christina. *In the Wake: On Blackness and Being.* Duke University Press, 2016.
Taussig, Michael. "Culture of Terror—Space of Death: Roger Casement's Putumayo Report and the Explanation of Torture." *Comparative Studies in Society and History* 26, no. 3 (1984): 467–497.
Williams, Raymond. *The Country and the City.* Chatto and Windus, 1973.
Williams, Raymond. "Ideas of Nature." In *Culture and Materialism: Selected Essays,* 67–85. Verso, 2005.
Yusoff, Kathryn. *Geologic Life: Inhuman Intimacies and the Geophysics of Race.* Duke University Press, 2024.

9

Disorder of Time in the Anticolonial Museum

Gabriel Giorgi

Translated by Nicolás Campisi

When we open *El museo de la bruma* (2019) by Chilean writer Galo Ghigliotto, we find the following image, a sort of strange geolocation:

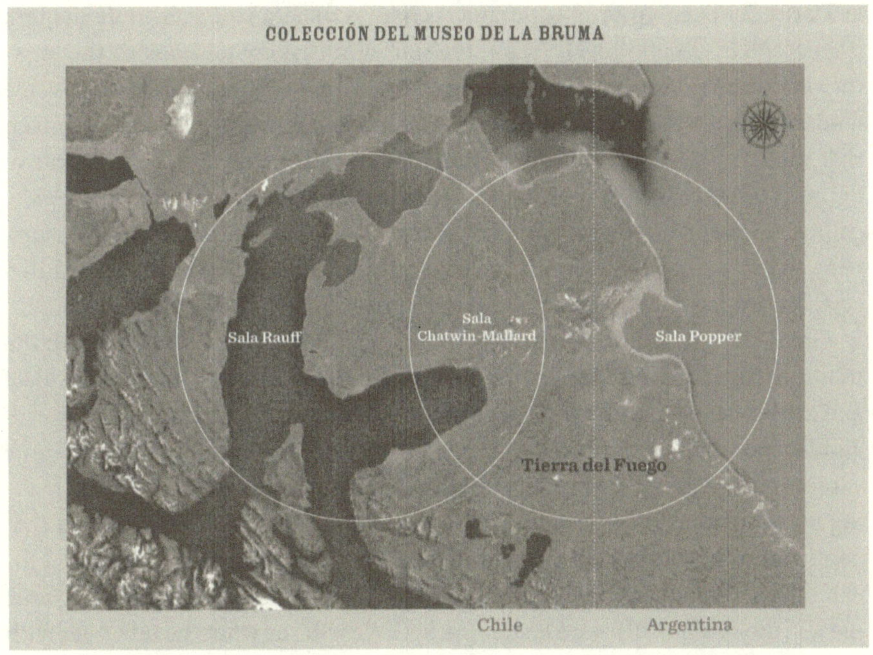

Figure 9.1. Galo Ghigliotto, geolocation image, *El museo de la bruma*, 2019. Courtesy of Laurel Libros.

Adapting Borges's "On Exactitude in Science" (1946) with very different consequences, here the museum coincides with the territory: the "Museum of Fog," as the book is entitled, divides Tierra del Fuego into three "rooms," each with its own name that links local and global histories. The book speculates on Tierra del Fuego as a territory irreducible to the Argentine or Chilean map and each country's modern history. It does narrate local stories about Patagonian and Fuegian colonization and how the construction of the modern Argentine and Chilean states is projected from the remote south, but it also presents a global historicity of violence that reaches into the present day. Ghigliotto's text geolocates this region to stamp the museum onto the territory: the museum, here, *is* the territory.

Ghigliotto's Museum of Fog is an imaginary institution whose catalog is included in the text. In an introduction called "Panel," as if the book were a curatorial text, we learn that the museum was founded toward the end of World War II (later explained as founded by "Dr. Morel and his wife, Faustine," thus ensuring its fantastic genealogy) and that it burned down in 2014, extinguishing its facilities and its permanent collections: "The inaccessibility of its location and the power of the winds of Tierra del Fuego allowed the flames to do their work impeccably" [La inaccesibilidad de su ubicación y la potencia de los vientos de la Tierra del Fuego permitieron que las llamas hicieran su trabajo de manera impeccable] (Ghigliotto 9).[1] Then, the museum that coincides with the territory is consumed by the very nature of that land: everything in this inaccessible landscape leads to disappearance. There remains, however, the catalog, given that the narrative "we" (and this is all readers will know about the narrative figure), claims to have found "in a small print shop in Porvenir the incomplete galley proofs of the first and only printed catalog of the museum" [En una pequeña imprenta en Porvenir las galeradas incompletas del primer y único catálogo impreso del museo] (9), material from which the book is composed.

Ghigliotto's text thus designs a fictional machine in which each piece in the museum becomes the index or fragment of a series of stories that accumulate and are juxtaposed under the figure of the exhibition; the series accrues stories and temporalities and mimics, through the use of the catalog, the museum's exhibition circuit. *El museo de la bruma* is a text installation in which reading is contiguous to seeing and writing to exhibiting; the book works on that continuum that links writing and installation, narrating and showing. This exhibition format is problematized in more than ironic ways, given that most pieces (though not all) were lost in the fire. We will see what the text does with that absence and with the gesture of showing the object or the material that is not there. For the moment, let us underline the general procedure of this narrative machine, that of situating the museum as a mechanism for accumulating, juxtaposing, and exhibiting diverse, largely heterogeneous temporalities and narrating what arrives from the south. Tierra del Fuego here becomes a sort of embryo of the twentieth century. A collection of objects, texts, and discards from which to trace the saga of Tierra del Fuego, both territorial and planetary: censuses, landscape paintings, chronologies of trials, lists of arrested people, gold harvesters, letters, and inventories of visions of Indigenous people. The museum seems to absorb everything; in everything, from the most obvious to the most insignificant, it finds an exhibition material that will also be a narrative seed.

El museo de la bruma narrates the twentieth century from the continent's edge, from the end of the world. However, the book mimics documentary and archival procedures while also using documents and archives that are indeed historical and subjects them to its fictional operations. At stake is a question about what we understand by narration in the Anthropocene. The reader is confronted with what the Museum of Fog unfolds with incomparable skill: the new disorder of time in which a heterogeneity of temporalities, agencies, and scales defy recognizable narrative formats, both literary and historical. In this disorder of time, multiple, nonsynchronous temporalities, human and nonhuman, Western and non-Western, press with new force on the frameworks that define the collective narratives of the regional, the national, and the global. The book represents an insurrection of times that challenge the frameworks of temporal intelligibility from which shared experiences are narrated: a museum as a narrative device at the very limits of narration.

Turning Territories into Temporalities

Let us return to the image with which the book opens. Ghigliotto's text makes the rooms coincide with the island's territory. The three rooms—Sala Popper, Sala Rauff, and Sala Chatwin-Mallard—refer to diverse historical and literary contexts that would be difficult to bring together in a museum. From the sinister Julio Popper, the aspirational and brutal ruler of the Argentine side of Tierra del Fuego and a central figure in the extermination of the Indigenous communities, to Walter Rauff of Nazi genocide infamy protected by the Chilean government who lived until his death in Punta Arenas, the Museum of Fog imagined by Ghigliotto is an archive conjugated around cycles of violence that defined the twentieth century in Latin America and around the globe. In the Chatwin-Mallard room, history gives way to literature since Chatwin was the "involuntary propellant of the current of illusory tourism" [Propulsor involuntario de la corriente del turismo ilusorio], and contemporary Mexican writer A. P. Mallard appears under the sign of the "collector of rarities" [coleccionista de rarezas] (13). This collection of stories, hardly groupable in a museum, is conjugated through the symbol of Patagonian fog.

What does it mean for a museum to coincide with the territory? Or, in other words, what does it mean to see and think of a territory as a museum? Unlike the coincidence between map and territory thought by Borges, who assembled a literal critique of representation and its ambitions, the coincidence between museum and territory leads in a different direction: it turns the territory into juxtaposed, accumulated, sedimented temporalities that do not allow sequencing by a narrative as an ordering principle but, on the contrary, are exhibited in their relative autonomy, multiple scales, and heterogeneous nature. The island, the landscape, and the space become blocks of time. Making the museum coincide with the territory produces a transcription in which the ground beneath one's feet is revealed as temporal blocks, layers of faster or slower speed, and active sediments made of time. The museum becomes a time machine, a time-activation machine, in which the territory reveals itself in its timeless spell of strata.

The notion of stratum, geological roots, refers to blocks of historicity, to processes and movements in which no element is subtracted from time; nothing, not even the rocky substratum of the earth, is outside of time. This deep geological time presents a radical challenge to common conceptions of history (Chakrabarty; Yusoff). Perhaps it is the already classic chapter 3 of *A Thousand Plateaus* (1980), "The Geology of Morality," where the decisive consequences of the notion of stratum are laid out. Gilles Deleuze and Félix Guattari mobilize this notion as a central argument around becoming, on which the whole project of *A Thousand Plateaus* revolves. The geology of morality is a bet on the division and multiplication of time in discontinuous and heterogeneous temporal lines whose model is neither human consciousness nor the life of bodies but the processes of the Earth and its stratigraphic modulation (Deleuze and Guattari). The stratum is a category of time and becoming, the heterogeneous, discontinuous, and diverse scales that make up a territory.[2]

I would like to situate the narrative procedure and the laboratory of temporalities that unfolds in *El museo de la bruma* around this conceptual operation. Transcribing territories into temporalities implies articulating a procedure in which formal tools are developed to account for a time irreducible to the human (especially to white, Westernized humanity, which functions as a measure of the collective histories of modern nation-states). In contexts of anticolonial disputes and environmental crises, nonhuman agents defy recognizable temporal frameworks. Territories made of multiple times, territories revealed in their temporal configuration, therein lies a decisive formal operation of this era. Territories, in other words, tell a multiplicity of stories that cannot be synthesized or unified in an encompassing narrative framework. Territories contain human stories of social, economic, and cultural processes such as colonization and the unfolding of capitalist accumulation but also nonhuman ones, environmental, mineral, animal, and climatic. Everything is revealed as historical (and for this very reason, the meaning of the historical is put into raging dispute, as Dipesh Chakrabarty contends); everything is made of time. We can say that such is the dictum of our era under the sign of the Anthropocene or the Capitalocene.[3]

El museo de la bruma provides an exceptional laboratory for that challenge. It is not alone in this task. A series of materials published mainly in recent years shows a remarkable aesthetic recurrence, works of diverse formats that turn territories into a discontinuous arc of temporalities. I am thinking, for example, of other text installations (and this is a key point: the influence of the installation over various forms of the written) such as *La compañía* (2019) by Verónica Gerber Bicecci and *Enciclopedia de cosas vivas y muertas: El lago de Texcoco* (2019) by Adriana Salazar Vélez, two works that resemble *El museo de la bruma* in their challenges to recognizable narrative formats (museum, encyclopedia, installation, and so forth). But I am also thinking of texts more anchored in the tradition of the novel, such as those of Colombian writer Juan Cárdenas, that interrogate in different ways the question of the Latin American landscape, which Cárdenas defines as "the scene of a crime," in a relentless sequence that goes from the travels of the naturalists in the nineteenth century to the current struggles against extractivism.[4] I am also thinking, in another formal extreme, of the series of "meetings" (*reuniones*) that Argentine artist Dani Zelko carries out with Indigenous activists and thinkers in which the political, historical, and aesthetic question of territory intersects with the forces and scales of the ancestral. The territory is thus revealed through poems that magnetize multiple temporalities.[5]

To turn territories into temporalities, to reveal the temporal fabric of that which we call territory or soil (and therefore region, Wallmapu, village, nation, city, planet, etc.), is the insistent question asked by a historical moment that has to face the fact that its condition of existence—capitalism and the colonial machine that undergirds it—leaves a geological trace, a material trace of planetary scale that will persist beyond our existence as a species. Faced with our geological footprint and the Earth's soil and subsoil (air and water) that reveal "time," the classic question arises: how do we narrate?

The Speech of Things

As I pointed out at the beginning, the catalog of this Museum of Fog is composed of lost pieces. We are left with the description but not the image, with few exceptions, of the exhibits. For this reason, the vast majority of the museum's pages present a similar format: the number of the piece, the description, and an empty box indicating that there was an object, but now there is only a blank space. The absent piece, the empty box, is the rule in this spectral museum.

This procedure allows the novel to include the most disparate materials as fragments belonging to different histories and temporalities while underlining the violent erasure perpetrated by colonial violence. The lost pieces, the empty tableaux that populate most of the pages, mobilize fiction to reconstruct the nonexistent archive of violence while alluding to its erasure as another dimension of colonial and genocidal violence. The missing pieces and empty frames thus function as echoes of violence that, if it cannot be reconstructed as a historical archive, resonates in the present, linking past and present outside the dominant frameworks of historicization, those of the nation-state or the global history of the contemporary. This dialectical nonimage, if you will, points to another idea of the museum, conceived as a sounding board in which colonial violence models the present from its very erasure and its spectral remains. If, as Shimrit Lee argues, "exhibition and colonialism went hand in hand" (104), by subtracting the image while leaving the mechanism of exhibition intact, *El museo de la bruma* combines the erasure of traces of the past with its spectral reverberation as the temporality of the colonial.

Let us look at "Pieza No. 160." The supposed photograph portrays a fragment of the Fuegian landscape with "stains that can be seen on the tundra" [manchas que se observan en la tundra], which are those of the skins of butchered dogs, skins piled up and abandoned (Ghigliotto 83). It is a scene of discarded debris of the living. The landscape is made with these remains. Because it records the junk and the rubble, the scene tells several stories and simultaneously speaks of this museum's historicity.[6] On the one hand, the confrontations between settlers and Indigenous people (the text, it should be added, abounds in chilling scenes of genocidal violence and unthinkable cruelty against Indigenous communities), resulting in this massacre of dogs. But at the same time, it is a scene displaying the commodification of colonial violence and its failure. The skins could have been sold, as they are mistaken for fox skins, but then discarded, supposedly because it was noticed that they would not find a suitable market. Appropriating land, commodifying human and nonhuman bodies, even miscalculating profit, all at the price of unspeakable violence, such is the postcard of conflicting temporalities that, like strata, colonial violence leaves in its wake.

PIEZA Nº 160

Matanza de perros yaganes

Las numerosas manchas que se observan sobre la tundra corresponden a los cuerpos de cientos de perros yaganes que fueron baleados en los enfrentamientos entre estancieros y selk'nam. En un principio, dado el parecido del perro yagán con el zorro patagónico, se había dado orden de contar, recolectar y apilar los cuerpos de estos animales para faenar las pieles, pero luego se decidió no proceder.

Inv. nº: 127
Material: fotografía, gelatina de plata sobre papel
Dimensiones: 11 x 13 cm
Sala: Ch-M

83

Figure 9.2. Galo Ghigliotto, "Pieza No. 160," *El museo de la bruma,* 2019. Courtesy of Laurel Libros.

The photograph is, then, that of the dispossession of Indigenous land and the residue of a failed commodification. It evokes an animal world and its intimate relation with Indigenous life. It is a figure of extinction that insists on the material persistence of those skins, transforming that landscape into a new permanence, a matter between the organic and the inorganic. Simultaneously, the photograph speaks of the massive death that came with colonization—human and nonhuman deaths—and indicates forms of temporal persistence that prevent us from thinking of extermination and even extinction as a simple erasure, a wiping off the map. Ultimately, the skins show the irrecusable, incontestable calling of the material memory of bodies and things.

The pieces in the museum's collection appear juxtaposed in temporal layers that resist being subsumed or synthesized in a general narrative. On the contrary, they function as strata, as temporal blocks that inscribe their own sequence, their immanent duration, their multiple scales that include social, economic, technological, and territorial configurations and that come with their own human and nonhuman perspectives and agencies.

This temporality radically disrupts an understanding of memory and its politics. As any good museum does, under the impulse of the exhibition, the Museum of Fog makes the objects speak; each piece tells its story because it has its own history recorded in the marks that time leaves on it and in the work of its material manufacture. Ghigliotto inverts the place of the subject: memory has to do, above all, with the objects' survival rather than the subjects' acts of memory. The object, accompanied by the gesture of archiving that turns it into a piece to be exhibited, bursts in and produces memory. This gesture is not so much that of the subject digging into the past as that of strata of the past returning as irruption, an interruption, a disorder of forms of intelligibility. The disorder of time comes above all from the persistence, both material and spectral (let us remember, of course, that we are dealing with empty paintings, the lack and emptiness of which are exhibited as a trace), of the materials, the objects, and their capacity to absorb the violence of which they were a part. The objects are instruments, witnesses, and remains. Ghigliotto zeros in on the colonial violence at the heart of the Latin American nation-states, preventing any fable that overcomes those acts of violence. *El museo de la bruma* disseminates the traces of that violence in active temporal strata. Since several dreams function as exhibited pieces, the oneiric world illuminates the continuation of this latent violence well into the present.

"Pieza No. 93" illustrates this dream of the thing as a museum material: the swastika lubricant tank. This piece invokes the type of historicity at play in the museum precisely, as a game, as an open field of possibilities. The piece is the Energina tank, which had the swastika as its logo, a product originating in an Anglo-Mexican company in 1908 that had nothing to do with Nazism or Germany and bore the swastika as a design gesture. How did this tank get to the museum? A copy of the Energina tank supposedly belonged to Alexander MacLennan, one of the mercenaries involved in the genocide of the Selk'nam people in Tierra del Fuego. In its global drift, the object knots heterogeneous temporalities in a commercial symbol in Mexico at the beginning of the twentieth century that would later condense the Holocaust and that contingently adheres to a protagonist of the Indigenous genocide in the south of the continent. Originally, the Energina tank had no relation either with Nazism or with the Indigenous genocide; it crosses and unites the two genocidal machines through its hazardous itinerary. It bears no causality, historical argument, or explanatory operation; it conveys pure coincidence, the contingency of pasts knotted in an insignificant object, a discard. This is the illuminating capacity of the objects, that of knotting temporalities, circuits, and sediments of time without necessarily turning them into signifiers of causality. In amalgamation, adhesion, and the juxtaposition of temporalities and senses, the object has the capacity to absorb the traces coming from the future and make them contiguous to the violence of the past. This is the method of activating times and senses at the heart of Ghigliotto's *El museo de la bruma*.

Here we could think of the theme of the fragment and its critique of unified history and totalizing meaning. In fact, this is a museum made of fragments and absence where the principle of unity and totalizing restoration is doubly betrayed; there is no unity or presence. But I believe something else, perhaps more interesting, is at stake here. I would rather emphasize the idea of stratum and how this notion allows us to think of temporalities or regimes of historicity. The stratum designates, above all, blocks of time that accumulate, overlap, and become contiguous without the promise of unification or totalization; rather, its contiguities are the occasion of friction, collision, and even seismic movements. The geological metaphor is apt precisely because the idea of the stratum entails locks of time, both human and nonhuman, natural and cultural. These stratified times, this stratigraphic time complicates any unified, homogeneous image of history, harboring heterochronies and diverse temporal sequences as well as relations with nonhuman temporalities, with material, organic, terrestrial, and cosmic processes that in this stratigraphic time claim citizenship status in how we think of the historical experience and the foci of the political.

In this disorder of time, non-Western and nonhuman temporalities exert a specific pressure that does not accommodate how Western natural history museums as well as national museums situate "pre-Columbian cultures" as a kind of surpassed anteriority. Piece number 12 provides a key moment for thinking about the place of the ancestral in the museum. It is an (unseen) illustration of an Indigenous legend. The legend cannot be dated, but it predates patriarchal rule and colonization. It comes from a mythical time, understood as the time of dialogue (and war) between humans and animals. A human leader, furious at losing one of her husbands to a tiger, decides to undertake the extermination of that species. She enlists the support of other female leaders, and all the tigers are eliminated. Before dying, the last tigress—note the centrality of the feminine in this narration—launches the curse and the prophecy on the Indigenous genocide. In the future, Western settlers will arrive to fulfill the curse of the tigress.

This legend posits a specific political historicity irreducible to Western forms of telling history. That anteriority, that mythical past, is the place where the Indigenous genocide takes place: the fury of a female leader in love and the revenge of the tigress that unleashes the end of the world. This temporality reappropriates the power of death—and with it, that of narration and its temporality—from the colonizers; the catastrophe comes from before. The politics at the heart of the disputes between humans and animals and between men and women, between species and genders, is what unfolds in the history of Indigenous extermination, the entry of the Indians into history under the figure of their disappearance. It is interesting to think of the place of that before, of that ancestry in *El museo de la bruma*, as a temporality that captures Indigenous historicity before colonization. That earlier history, that past of the past, that outside-the-archive, that which the archive cannot contain, is inscribed as a narrative perspective and an ancestral temporality in the museum's temporal disorder.

The Direction of Time

Toward the end of *El museo de la bruma*, we find Piece number 703, titled "The Real Reality." It is a text written in the museum visitors' book and incorporated as an artwork. This museum that absorbs everything also includes the reflection of a visitor, a spectatorship text that becomes a piece. It is a key text because it reflects on the fog that gives the museum its name and the book its title and considers the general selection and archiving procedure the book carries out. "Reality," its anonymous author says, "is a fog dotted with diffuse images: they do not become drawings, but strokes, perhaps dots" [La realidad es una bruma salpicada de imágenes difusas: no alcanzan a ser dibujos, sino trazos, acaso puntos] (Ghigliotto 288). Let us note the question of the exhibition where the fog is the common denominator of perception; the museum is, then, the repertoire of these fragments that emerge, a light on these brief interruptions about disorientation. This phenomenology of the half-light, of diffuse perception, where the clues are fragmentary, this darkening where nothing is ever seen definitively and totally, is resolved in temporal terms.

> It is hard to live among fragments. It is the constancy of living among ruins, the ruins of something that was or will be. We cannot know because this is how existence functions: stars that explode or melt to create planets or black holes, materials that are constantly traveling toward destruction or creation: we only see the current, the flow of particles, but not their direction.
>
> Es duro vivir entre fragmentos. Es la constancia de vivir entre ruinas, las ruinas de algo que fue, o será. No podemos saber, porque así funciona la existencia: estrellas que explotan o se funden para crear planetas o agujeros negros, materiales que están viajando constantemente hacia la destrucción o la creación: sólo vemos la corriente, el flujo de partículas, pero no su dirección. (Ghigliotto 288)

Here, I want to suggest, we have a key to the museum, to its spectral existence and its returns: an open collection of fragments whose temporality is uncertain, whose place in historical sequences is indeterminate or indeterminable. Instead of ordering the past in relation to the present, the museum verifies the impossibility of distinguishing between past and future, where the traces, the remains, and the marks of historical experience remain irreducible to a linear historicity. It is the museum of the "ancestral catastrophe" in the sense given by Elizabeth Povinelli, who thinks of colonial expansion as an event that, lodged in the past, does not cease to recur in the present, as in a kind of systematic iteration that operates as an unsurmountable condition of the modern (3).

It is fundamental for the text to represent and think about this indeterminacy through reference to physical and cosmological knowledge: matter and speed, stars and planets. The historicity of these fragments and material traces (a museum is ultimately nothing more than that, as a repertory of loose materials; Ghigliotto's text highlights this through the object in absentia) can never be reduced to their cultural, socioeconomic, and historical dimensions, to their human uses. The pieces are traversed by the deep time of the matter they are made of and the existence of an incommensurable scale for the eye of the present (stars, planets, and black holes). Precisely from that scale comes each object's indeterminacy. What is the direction of time? We must interrogate the flow of particles, that material, geologic, and cosmic agency that now directly intersects the time of culture and politics. This is where the question of ruin—or "rubble" in Gastón Gordillo's sense—comes to the fore: instances of interrogation of the archive and time in which the nonhuman exerts a new pressure on previous frameworks of legibility.

The fluctuation between past and future, between the past of the past and the futures and nonfutures (and, clearly, between the ancestral and the future) shapes the times we live in. To turn territories into temporalities is to fissure the time of modernity, its modulations of historical experience, and its dependence on the nation-state as a monopoly of collective temporalities. To imagine other histories and other ways of inhabiting the planet, such is best traced in the narrative laboratories of our present.

Notes

1 Unless otherwise indicated, all English translations are the translator's own.

2 As Deleuze and Guattari propose in *A Thousand Plateaus,* anticipating the contemporary geological turn and the centrality of deep time, the concept of the stratum brings a perspective in which minerals and rocks offer the model of the material composition of existence. The stratum, defined by its "double articulation" between matter and expression, allows Deleuze and Guattari to formulate the tension between territorialization and deterritorialization—and thus the very notion of becoming—without making any concession to teleological notions of evolution, that is, without the emergence of a synthesis or new formation that subsumes previous processes and movements (Deleuze and Guattari 40). In this sense, the geological modeling of time allows for an analytical framework that accounts for the emergence of new temporal formations without losing sight of sediments and layers that are not metabolized by the previous stratum. At the same time, by inscribing geological time in the analytics of time, the notion of stratum dislocates any biocentric or humancentric model of temporality. It is not culture or the living body but rocks and minerals that provide the key point of view for thinking time. The stratum is the index of interrupted temporalities, heterogeneous formations, and divergent scales. They maintain mutual interrelationships by articulation or friction, avoiding the models of temporality inspired by organic life. While the notion of stratum offers the model of how matter finds its form—in geological, organic, and human configurations—it always maintains an active threshold of deterritorialization that produces the "aberrant movements" that make possible new becomings (Lapoujade). Interestingly, Deleuze and Guattari's conception of stratum revolves around resonance and vibration, an aural semiotics that speaks of alternative ideas of expression and communication, placing materiality at the center of meaning rather than the mind, language, perspective, or even the self (Buchanan). The stratum thus articulates the possibility of different velocities and scales that resist any unified or linear conception of time. It simultaneously inscribes the relation to the nonliving and the nonhuman without subsuming them into a biocentric or humancentric temporality; it differs, in such a sense, from Reinhart Koselleck's historiographical approach to the notion of stratum as entirely focused on human social formations.

3 Debates on (neo)extractivism, climate change, Anthropocene/Capitalocene, and bio/cosmopolitics are confronted with the fact that inherited conceptions of historical experience along with narrative form and its temporal frames are challenged by the inescapable pressure of nonhuman (or nonanthropocentric) scales and perspectives and by the new prevalence of the nonliving, the extinct, and the inorganic. As named by François Hartog, we are living in a new "regime of historicity" in which a multiplicity of temporal scales is juxtaposed and enter into friction with each other, making any progressive and teleological conception of history unfeasible. Likewise, as Chakrabarty has noted, the intertwining of historical time and deep or geological time (what he would later call "the planetary") puts new pressure on our conceptions of historicity, so deeply shaped by anthropocentric and human-social scales. The temporal seism, profusely observed from a historiographic perspective, is part of a broader shift that defines many of the key issues within contemporary aesthetics, the need to rethink and rework frameworks of temporal intelligibility that help us to understand the current crisis and to orient our collective action.

4 The cited passage appears in Juan Cárdenas's unpublished manuscript, "Obscenidad del paisaje y pintura de viaje," which he shared via personal communication.

5 An example of these *reuniones* as a listening experiment on the territory and Indigenous politics is Dani Zelko and Soraya Maicoño's *Pewma Ull: El sueño del sonido,* at Reunión, https://reunionreunion.com/el-sueno-del-sonido.

6 See Gordillo for a theory of rubble "as a conceptual figure that can help us understand the ruptured multiplicity that is constitutive of all geographies as they are produced, destroyed, and remade" (2).

Works Cited

Buchanan, Ian. *Assemblage Theory and Method.* Bloomsbury, 2021.

Chakrabarty, Dipesh. "The Climate of History: Four Theses." *Critical Inquiry* 35, no. 2 (2009): 197–222.

Deleuze, Gilles, and Félix Guattari. *A Thousand Plateaus: Capitalism and Schizophrenia.* Translated by Brian Massumi. University of Minnesota Press, 1987.

Ghigliotto, Galo. *El museo de la bruma.* Laurel, 2019.

Gordillo, Gastón R. *Rubble: The Afterlife of Destruction.* Duke University Press, 2014.

Hartog, François. *Chronos: L'Occident aux prises avec le Temps.* Gallimard, 2020.

Koselleck, Reinhart. *Sediments of Time: On Possible Histories.* Stanford University Press, 2018.

Lapoujade, David. *Deleuze, les mouvements aberrants.* Editions de Minuit, 2014.

Lee, Shimrit. *Decolonize Museums.* OR Books, 2022.

Povinelli, Elizabeth. *Between Gaia and Ground: Four Axioms of Existence and the Ancestral Catastrophe of Late Liberalism.* Duke University Press, 2021.

Yusoff, Kathryn. *A Billion Black Anthropocenes or None.* University of Minnesota Press, 2019.

10

Emilio Renart's New Sense of Space

Florencia Malbrán

Emilio Renart, an influential figure in Argentine art during the 1960s, has reemerged with new relevance today. Renart committed his artistic career to radical experimentation, explored novel themes and materials, and devoted himself, above all, to unfolding the fundamental concept of "integration." His intention to integrate humankind and the universe prefigures recent debates proposing continuity, interconnection, and interdependence between humans and nature.

Renart produced sculptures and paintings and dedicated himself to the study of creativity, writing a book on the subject and conducting workshops that established a direct relationship between the participants and the creative act in order to influence first the individual and then the social fabric. He became known for his *Bio-Cosmos* series, composed of five major works representing his concept of "integralism." In what follows, I focus on the works in this series to revisit how the past can help us recognize the potential of present debates about our planetary condition, which is beset by crises and vertiginous changes. I examine how Renart's work was perceived in his time and from what perspective we can observe it today.

Renart began studying art in Mendoza, far from Buenos Aires, the country's cultural center, and moved to the capital when he came of age. In 1962 he exhibited his *Integralismo Bio-Cosmos n 1,* the first artwork of the *Bio-Cosmos* series that would garner him recognition on the national art scene. He added a short text that he wrote for the exhibition of this work.

Renart began his sentences—almost all of them—with the idiom "face to face with" [frente a], in order to mark his opposition to the poles and divisions that structure an understanding of the world and that we can recognize as the binary oppositions employed in Western thought, among them passivity/

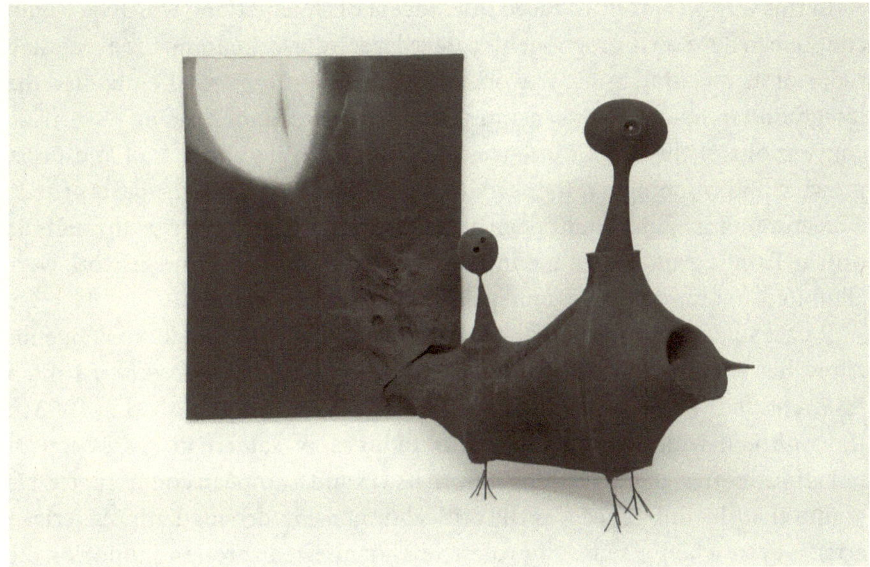

Figure 10.1. Emilio Renart, *Integralismo Bio-Cosmos n 1*, 1962. Photograph by Viviana Gil, Courtesy of Museo de Arte Moderno de Buenos Aires © Emilio Renart Estate.

activity, culture/nature, and feminine/masculine. Renart opens his text *Renart expone* with "Face to face with a reality that tends to divide efforts" [frente a la realidad que tiende a subdividir los esfuerzos].[1] He goes on to acknowledge "fluctuating states" and marks the connections between the "absolute," which would be the planet or *cosmos*, and "individual integrality," humanity or *human biology*. Renart thus coined the term "bio-cosmos." He concludes *Renart expone* with a resolution in favor of integralism.

How did the work on display express this integrative motto? A figure emerged from a brown canvas; the painting opened into space and became an organism that advanced across the gallery floor and stood on three legs. Renart had concocted a creature that challenged the viewer. "Face to face with the flat plane, or volume or stark lines," he had decided instead to integrate painting and sculpture into a thing, a monster, a strange animal, a robot, or even perhaps a spaceship.

The canvas, with its craters, had a texture recalling the surface of planet Earth, which Renart had achieved by manipulating materials: sand, plaster, and sawdust in shades of brown. Renart gave his organism a "head," a bulging appendage with a glass eye. Underneath, in the belly or chest, he fashioned an opening that allowed viewers to observe the creature's internal anatomy.

In this way, Renart introduced the concept of "integration," which he would continue to develop throughout his career, layer by layer, refining the expanding scope of its meaning in his artworks and texts. *Bio-Cosmos n 1* embodied the integration of media, that is, painting and sculpture, undermining the artistic conventions of the time. Moreover, the work activated both wall and floor, growing and reaching out to the viewer, who, standing amid the parts of such a creature, in its guts, could peer further into the beast's interior through its orifice. Painting and sculpture, interior and exterior, viewer and artwork were all united. Yet integralism would gain deeper semantic strata.

Rafael Squirru, the former director of the Museo de Arte Moderno in Buenos Aires, has acknowledged the relevance of *Integralismo Bio-Cosmos n 1* in an essay entitled "Pop Art or the Art of Things," which was published in 1963 in the continent-wide magazine *Americas.* In his essay, Squirru discusses pop art in Latin America, distinguishing it from its US and European counterparts. He argues that the difference was the critical dimension driving Latin American artists, even when a sense of protest was manifested through humor or the uncanny. In Latin America, artists did not limit themselves to the mere appropriation of objects of consumption but instead included those very objects to annihilate them, to question their very existence. Objects became protagonists, targets of criticism, so Squirru preferred the term "the art of things" to "pop art," which he suggested should be dismissed. He assessed the practices of several artists, such as Marisol, Marta Minujín, Claes Oldenburg, Dalila Puzzovio, and Jean Tinguely. However, Squirru gave Renart a prominent space within this group.[2] His essay in *Americas* is illustrated first with a photograph of *Integralismo Bio-Cosmos n 1,* emphasizing its importance, and then an image of *Integralismo Bio-Cosmos n 2.* Squirru contends, "The Art of Things is an art of protest" and that works by these artists were a "cry of rebellion against things," a protest against consumerism and alienation, against the exploitation of the planet. His analysis suggests that these artists recognized the intricate connections between living beings and their environment, thereby questioning the notion of "nature" as distinct from humanity.

Renart exhibited his *Integralismo Bio-Cosmos n 2* in 1963 at the Museo Nacional de Bellas Artes in Buenos Aires. From a canvas hanging on the museum wall emerged a figure larger than the one in his previous work. This figure stood on multiple legs and had a neck crowned with a head with one eye. Additionally, it had a trunk and an elongation at one end of its body. The rough surfaces of the work resembled the crust of planets or the inner walls of human or animal organs. The following year, Renart exhibited *Integralismo Bio-Cosmos n 1* and *n 2* together at the Museo de Arte Moderno in Buenos Aires, combining them into a single work.

Renart became an original figure within the Argentine art scene. His art did not fit neatly into pop art, as Squirru noted. His works, so profoundly original, attracted critics; however, these critics struggled to explain them through the frameworks of artistic movements of the time. Looking at *Integralismo Bio-Cosmos n 1,* the writer Manuel Mujica Láinez remarked, "Renart presents a work that is probably the strangest we have seen in Buenos Aires" [Renart presenta una obra que será probablemente lo más extraño que hemos visto en Buenos Aires]. He further emphasized its rarity by noting, "Both the pictorial and sculptural aspects are perfectly achieved, and this unusual fusion constitutes a discovery that will undoubtedly be much imitated and perhaps will inspire valuable creations" [Tanto lo pictórico como lo escultórico están perfectamente logrados y su insólita fusión constituye un hallazgo que será sin duda muy imitado y acaso originará creaciones valiosas] (in Lebenglik 25). In the same vein, José Gómez-Sicre, chief curator at the Organization of American States in Washington, DC, where Renart exhibited in 1965, posited, "Renart has excelled with three-dimensional compositions or 'monsters' that are a cross between strange mechanisms from outer space and weird animals with tentacles and enormous eyes, which seem straight from the pages of science fiction." Renart's contemporaries noticed that the artist departed from the norm by combining disjunctive media and techniques. They also realized that Renart was exploring the interface between the organic and the technological and witnessed fantastic beings that were testing the limits of reason. The interpretations that Renart's art generated even led him to become associated with surrealism. So was the understanding of critics Hugo Parpagnoli and Aldo Pellegrini. In 1967, when Renart was invited to represent Argentina at the São Paulo Biennial, Parpagnoli wrote in his catalog essay that Renart's works "gives a new impetus to contemporary surrealism" [da un renovado brío al surrealismo contemporáneo] (in Giesso 3). That same year, Aldo Pellegrini included Renart in the group exhibition *Surrealism in Argentina,* which he organized at the Instituto Torcuato Di Tella.

Renart presented *Integralismo Bio-Cosmos n 4* in 1965.[3] This work consisted of two canvases suspended from the wall from which a third form emerged, concave, projected forward, and resting on a pole or foot. A light flickered and illuminated the work. On the sides, the canvases on the wall had a roughened texture, again evoking the surface of the earth—a rugged expanse, a rift—or an organ; they resembled the alveoli found in the bronchioles of the lungs that enable breathing, that is, life. In the center, the concave form was articulated with resin spheres of various sizes. Oscillating between figuration and abstraction, this work displayed "spheres as symbols of the intellect for liberated energies that would return to the universe to generate life again and again" [esferas como símbolos del intelecto por energías liberadas, que retornarían al universo para generar vida una y otra vez], according to Silvia de Ambrosini, curator of the Argentine contribution to the 1967 São Paulo Biennial (19). The term "liberated energies" could then be displaced and used to define "the vital force of life," that "transversal force that cuts across and reconnects previously segregated species, categories, and domains," as Rosi Braidotti puts it ("Posthuman" 686). The integration proposed by Renart can go beyond the individual and include the nonhuman.

This work, *Integralismo Bio-Cosmos n 4,* was exhibited at the Instituto Torcuato Di Tella as part of the annual prize organized by that institution. The entry dedicated to Renart in the prize catalog included a text and a portrait of the artist. Renart's photograph merits special attention. It was intervened by vectors, lines that crossed diagonally over a part of the artist's own image. Renart was integrating his face into the fabric of the world, blurring all bodily boundaries. If Renart's face was connected to the planetary fiber, was there an inside and an outside? Was this photograph an individual portrait or rather the expression of a body without organs?[4]

The term "integralism" came to mean all matter, regardless of its place in the universe. Renart wrote about "cosmogonic harmony," a harmony that could be formed by the interrelationships among human beings, organisms, and technologies. He states, "I believe that nature, in its continual change, gives us the guidelines of what to do, since we are part of it and not just spectators" [creo que la naturaleza, en su modificación continua, nos da la pauta de lo que debemos hacer, ya que somos parte de ella y no meros espectadores] (*Multimágenes*). Therefore, it is possible to understand Renart's integrative intention from the present as a prefiguration of the critique of humanism. It is also possible to consider the way in which Renart invited his viewers to embrace our planetary condition.

The humanities—built on the ideal of "man" as the universal measure of humanity—have been under intense scrutiny for at least four decades. The notion of humans as the measure of all things, the faith in reason based on a binary structure of thought, and the teleological vision of progress have been questioned. Posthuman theories have deepened these inquiries and redefined subjectivity as an expanded and relational entity. Braidotti explains, "A more complex vision of the subject is introduced in a materialist process ontology that foregrounds an open, relational self-other entity framed by embodiment, sexuality, affectivity, empathy, and desire as core qualities" ("Posthuman" 681). Posthumanism presents a nature-culture continuum that replaces the binary oppositions of Western thought. This continuum implies a turn toward a monistic ontology: we no longer see oppositions (such as the dualisms of human/

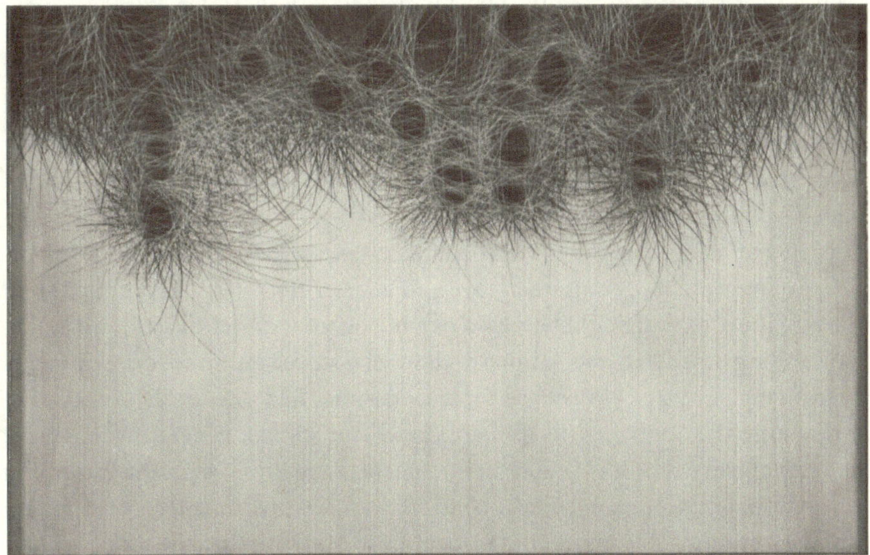

Figure 10.2. Emilio Renart, *Dibujo No. 21,* 1965. Colección Eduardo F. Costantini. © Emilio Renart Estate.

cosmos, materialism/idealism, human domination/dominated animality) but rather extensions and relations woven into and across various surfaces. In Renart's words, these are "integrations."

The emphasis is now on living matter. Contemporary debates on subjectivity point out that life is no longer the exclusive prerogative of humans. Instead, there are nonhierarchical relationships between animals, the earth, and other beings; all participate in a symbiotic system of codependence and coproduction. Even in Renart's drawings one can find an expression of these relationships, transitions, and hybridization processes in which we are all involved.

A dark, undulating shape runs along the entire top edge of *Drawing No. 21* (1965). This shape has no center, and there is no axial symmetry at all in the work. Within the black mass, vectors—are they filaments? hairs? rays?—grow and accumulate, intercepting each other or leaving gaps. Renart drew "a continuous, self-vibrating region of intensities whose development avoids any orientation towards a culmination point or an external end" (Deleuze and Guattari 2). This famous phrase, which I borrow from Gilles Deleuze and Felix Guattari, aptly describes the fields and extensions that Renart traced.

Bio-Cosmos n 3 (1964) consisted of three parts. First, in the center, a large sculpture extended along the floor and wall. On the wall, Renart placed an oval shape with an opening or slit that resembled a vagina. On the floor, he placed a heap of wires that appeared to be a female mount of Venus and pubic hair. Then two other lateral parts or volumes developed vertically, also alluding to organic or sexual cavities.

The sexual tone was clear. It responded to the climate of an era in which the experience of sexuality was lived with greater freedom. In Argentina, the economic prosperity that accumulated since the post–World War II period—which saw the production of cars, steel, and fuel, as well as the linkage of the concepts of "democracy" and "development"—was accompanied by the emergence of new media outlets, the growth of universities, and the propagation in them of innovative thought based on psychoanalysis, existentialism, and Marxism, among other transformations of the social structure. These developments precipitated a crisis in the conservative ideas that had previously prevailed in education. The contraceptive pill appeared and spread, generating extensive debates on sexual morality and birth control. Sexuality was in the midst of a revolution, and in Renart's *Bio-Cosmos* we can observe an appropriation and resignification of this burgeoning imaginary. Moreover, this work of art was presented at the Instituto Torcuato Di Tella, an epicenter of the avant-garde and a showcase for new lifestyles that were characterized by a self-sufficient and rebellious sensibility that stood in contrast to the arguably more restrained and traditional order of previous decades.

Renart's interest in biology accompanied his focus on the space race. The exploration of the cosmos in the 1960s appeared to indicate an unlimited acceleration of progress. Satellites were launched, spacecraft departed from Earth's orbit, and mass media disseminated information about meteorites, planets, galaxies, and the composition of the atmosphere. The advent of a new knowledge of space ignited the imagination of artists. Renart shared the fascination caused by the novel images of the universe with other prominent Argentinean artists, such as Raquel Forner, Antonio Berni, and León Ferrari. Forner, for instance, dedicated herself to creating successive series of paintings on the theme of space beginning in 1957, the year Sputnik was launched. She also traveled to the NASA Space Center in Houston in 1974. Berni had an astronaut greet his very popular and endearing character Juanito Laguna from the window of his spaceship as he flew through the sky looking down on the slums, as seen in Berni's collage *El cosmonauta saluda a Juanito a su paso sobre el bañado de Flores* (The Cosmonaut Salutes Juanito as He Passes over the Flores Swamp, 1961). Ferrari created a spherical sculpture entitled *Gagarin* (1961), evoking the orbits of the *eponymous* Soviet pilot on his journey into space. However, Renart's enthusiasm for exploring the universe did not entirely align with the inflated belief in progress, which saw humans as the ultimate artificers of the cosmos.

Renart's sensitivity to humanity's leap into the unknown led the popular-science magazine *2001* to dedicate a feature to him in 1969. The editors of this Spanish-language magazine, which specialized in the space age, were enthralled by Renart's works because they saw in them an anticipation of the conquest of the moon, which took place the same year. They asked Renart, "To what does your interest in the cosmos, these lunar landscapes, this personal passion for flying saucers correspond?" [¿A qué responde ese interés por el cosmos, esos paisajes lunares, esa pasión personal por los platos voladores?]. Renart replied,

> I made lunar landscapes and I make biocosmic monsters because I am not satisfied with the limitations of life, and I need a space greater than that of the Earth. I believe in UFOs violently, without hesitation. In the worst case, if they do not exist, then they are a human creation, and that makes us important. Human imagination, creativity, transform us.
>
> Hice paisajes lunares y hago monstruos biocósmicos porque no estoy satisfecho con las limitaciones de la vida, y necesito un espacio mayor que el de la tierra. En los OVNIS creo violentamente, sin vacilación. En el peor de los casos, si no existieran son una creación del hombre, y eso ya los hace importantes. La imaginación humana, la creación, nos modifica. (Renart, "Alunizar desde la Tierra" 57)

In this answer Renart emphasized the value of creation. It is already evident here that the subject of creativity would occupy him once the *Bio-Cosmos* series was completed. He would then devote himself to teaching, to writing a book entitled *Creatividad* (Creativity, 1987), and to conducting massive workshops with artists and the general public.

In 1967, *Bio-Cosmos n 5* was exhibited in the São Paulo Biennial. Renart devised a substantial installation, commencing with a sizable nucleus constituted by a multitude of spheres, exceeding a dozen, suspended from the wall. From the nucleus, a chainlike structure emerged, comprising a series of spheres aligned and resting on a bed of sand. The work advanced through the space until it reached a protrusion on the floor. The spheres, crafted from polyester resin, housed a light bulb within; the work was both electrified in a literal and a metaphorical sense, imbued with a vibrant energy. Renart was a highly skilled practitioner of plastic resin. The previous year, he had participated in the group exhibition *Plástica con plásticos* (Plastic with Plastics) at the Museo Nacional de Bellas Artes. In preparing their works for the exhibition, Renart and the other participating artists received technical guidance as well as tools and materials provided by the leading Argentine companies involved in the plastics industry.

Bio-Cosmos n 5 was supposed to be sent to Buenos Aires after the São Paulo Biennial. But it never arrived in its entirety. Shipping problems caused some of its components to go astray. This was not the first time Renart suffered losses. Of the five works comprising the *Bio-Cosmos* series, key in Renart's conception of integralism, only two have survived. Just *Bio-Cosmos n 1,* housed in the Museo de Arte Moderno in Buenos Aires, and a portion of *n 3* remain. Following its exhibition at the Instituto Torcuato Di Tella, the latter work was displayed at the Mar del Plata Salon, where it was awarded the acquisition prize and thus became part of the collection of the Museo Provincial de Bellas Artes de La Plata. But due to poor conditions at the storage facility of that museum, the work deteriorated considerably, and only the central part could be preserved. In São Paulo, along with his installation, Renart showed a suite of drawings acquired by a museum in Rhode Island and lost in transit to the United States. These losses, coupled with years of tireless work, took their toll on the artist. Renart then shifted his career to concentrate on the pedagogical power of art.[5]

All the works that make up the *Bio-Cosmos* series are hybrids; they are crossings between painting and sculpture, between wall and floor, and between human life and outer space. Renart seems to question the standard that upholds humankind as a universal measure excluding all other existing organisms that do not conform to this measure. Bios and cosmos are one, fused, integrated. Human beings are not exceptional, and relational capacity extends to the entire planet through continuous and complex negotiation with dominant norms. Living matter—that "process ontology that interacts in complex ways with social, psychic and natural environments, producing multiple ecologies of belonging" (Braidotti, "Posthuman" 686)—has a planetary dimension. This new planetarity is embedded in the current context of severe ecological crisis and social struggles to achieve a horizontality that cuts across the old hierarchies that have marginalized other modes of existence. "Planetarity," Mary Louise Pratt explains, "resonates above all with what I have been calling the crisis of futurity linked to climate change and the impending social catastrophe" (10).[6]

It is also interesting to note that theorists engaged in the critique of gender norms have equated posthuman thought with postgender, even proposing a kinship between monsters, aliens, and political subjects that oppose binarisms in alliance against the capitalist, patriarchal, and colonial order. In this sense, La Chola Poblete, an artist who explores hybridity and changes in the formulation of sexual identity, has mentioned Renart as one of their referents. In 2023 La Chola Poblete commented on their interest in the "eroticism" of Renart's drawings and in the interweaving between the drawings and the "organic sculptures" and suggested that Renart invites us to "think about the mutant thing of changing over time."[7]

The widespread impact of Donna Haraway's "Cyborg Manifesto" can also be linked to the renewed interest in the figure of the alien and the need to conceive other modes of planetary existence. Haraway distances herself from the "border war" of "Western science and politics" and proposes instead clinging to the cyborg, understood as "machine and organism," a "hybrid" being, a postgender, postrace creature that modifies and displaces the boundary between living bodies and machines, between inside and outside (149). Today, Haraway's ideas are being incorporated into countless exhibitions as curators worldwide question the notion that the human is born whole and remains stable throughout life. The cyborg illuminates the transition from single identity to hybrid subjectivity, one that would be fluid, intersectional, and networked, without a defined locus of origin and without an endpoint.

The cyborg is as uncanny as attractive; in its strangeness, it epitomizes the potential that the blurring of established boundaries implies, and it presents itself as a model of promising mobility. Indeed, many contemporary artists are exploring a liberation of what is considered human, blurring and transgressing the limits of the body by emphasizing the continuous and multiform characteristics of life. In anticipation of these transgressions, Renart's *Bio-Cosmos* figures, like cyborgs, already eroded the boundaries that separate differentiated regimes of nature and culture, animate and inanimate, human and animal, organic and technological. It is fitting that a Renart retrospective was titled *Alienígena;* it included more than two hundred works produced by the artist between 1962 and 1989, the year of Renart's last solo exhibition, shortly before his death in 1991. Sebastián Vidal Mackinson, the curator of this comprehensive retrospective, observes that the term "alien" refers to "the strange status of a practice that operated in the Buenos Aires art scene in a powerful way, envisaging problems that have emerged strongly in recent years" [la condición extraña de una práctica que operó en la escena artística porteña de manera contundente y se adelantó a problemas que han hecho su fuerte aparición en los últimos años] (*Alienígena*).

This exhibition was held in 2024 at the Colección de Arte Amalia Lacroze de Fortabat in Buenos Aires. The recent opening of an archive of documents kept by the artist allows for a new understanding of Renart's work. The firsthand encounter with the works of art, as well as the analysis with new historical sources, opens new approaches to integralism and relevant research questions that intertwine Renart's art with contemporaneity. These new approaches are crucial to decolonizing art history and including global contributions. Indeed, Renart's work intervenes in the international history of art by offering unique perspectives on our complex relations with the environment, nature, and the threats to our planet.

Renart's work helps us to see the fabric of planetary ties that sometimes tighten and sometimes seem to fork, to loosen. His work is speaking to the now. Renart did not seamlessly anticipate the debates over posthumanism and planetarity. Making the past present through his art is not without tension, and even Renart himself sometimes commented on his longing for a type of "humanism." Yet, despite the presence of frictions, an analysis of his oeuvre from the perspective of contemporaneity can be productive. A positive insight is revealed: "We" have significant differences, but we are on this planet together as a species.[8] Renart railed against "thanatic models" and called instead for "playing with Eros." He quipped, "My choice is quite clear. The foundations of integralism do nothing more than give theoretical flesh to this choice for life. The five *Bio-Cosmos* works have exalted the generation of matter, at the moment of pure and primary creation" (in Pérez). Let us collectively create a new planet.

Notes

1 Unless otherwise indicated, all translations are my own.

2 Squirru also highlights Renart's art in his article "Spectrum of Styles in Latin America," published in *Art in America* in February 1964.

3 Before, in 1964, Renart had exhibited *Integralismo Bio-Cosmos n 3,* which I address later in this essay.

4 I am referring here to Deleuze and Guattari, authors whose philosophy, in the words of Argentine theorist Luz Horne, represents an "epistemological opening" that allows us to imagine ways of overcoming the world's political and ecological crises. Horne notes that "although scientists have believed they are capable of sharply separating all that is nature from humanity, or the rational from the intuitive, the work of science involves hybrid objects (human and nonhuman) that defy separation" [aunque los científicos se hayan creído capaces de separar de modo tajante todo aquello que es naturaleza de lo que es humanidad o aquello que es racional de lo que es intuitivo, el trabajo de la ciencia involucra objetos híbridos (humanos y no humanos) que desafían la partición] (27). In *Futuros menores,* Horne explores modes in which Latin American literature and art generated alternative lines of thinking to those of twentieth-century science, thus enabling different, more sustainable ways of understanding time and space.

5 Curator Sebastián Vidal Mackinson studied the *Bio-Cosmos* series as well as Renart's pedagogical work, including the organization of collective workshops to explore creativity. Vidal Mackinson explains, "Tired of the 'hypercompetition in the art world,' Renart began in 1968 a transition to become what he termed a 'social artist,' positioning himself at the fold or intersection of art and education." He also notes that "from that point on, Renart started to conceptualize the term 'creativity' with greater rigor" [Cansado de la hipercompetencia (*sic*), inició una transición hacia lo que denominó "artista social" y se situó en el pliegue de la relación entre el arte y la educación . . . A

partir de allí fue conceptualizando con mayor rigor el término *creatividad*] (*Alienígena*). Curator Javier Villa, who organized the exhibition *Constancia de especie: The Archive of Emilio Renart,* emphasizes Renart's goal of "finding a tool for social healing through coexistence, exploring creativity, and recognizing the multiplicity of the individual within the collective" [encontrar una herramienta para la sanación social con la convivencia, la exploración de la creatividad, la multiplicidad de lo individual dentro de lo grupal] The show "*Constancia de especie.* The Archive of Emilio Renart" was on view at the Galería del Infinito in Buenos Aires in 2024.

6 Terry Smith also links planetarity to destruction. When imagining Gayatri Chakravorty Spivak's notion of the "planetary" beyond comparative literature (as posited in Spivak's seminal book *Death of a Discipline*), Smith examines ways in which artists may attempt a "weaving of connectivities between states and planes" to mark place and trace constellations in order to slow down extinction (Smith 171).

7 La Chola Poblete, email to the author, July 25, 2023.

8 I am following Braidotti, who looks into transfigured futures created through the power of the collective subject "we," that is, "the dwellers of this planet at this point in time" ("We Are in This Together" 261).

Works Cited

Ambrosini, Silvia de. *Coincidencias en espacios-tiempos del arte.* Nobuko, 2004.

Braidotti, Rosi. "Posthuman Feminist Theory." In *Oxford Handbook of Feminist Theory,* edited by Lisa Disch and Mary Hawkesworth, 673–698. Oxford University Press, 2016.

Braidotti, Rosi. "We Are in This Together: Posthuman Times and Affirmative Ethics." In *Goshka Macuga: Before the Beginning and after the End,* edited by Mario Mainetti, 255–261. Fondazione Prada, 2016.

Deleuze, Gilles, and Felix Guattari. *A Thousand Plateaus: Capitalism and Schizophrenia.* Translated by Brian Massumi. University of Minnesota Press, 2005.

Haraway, Donna. "Cyborg Manifesto: Science, Technology, and Socialist Feminism in the Late 20th Century." In Haraway, *Simians, Cyborgs, and Women: The Reinvention of Nature,* 149–181. Routledge, 1991.

Horne, Luz. *Futuros menores: Filosofías del tiempo y arquitecturas del mundo desde Brasil.* Ediciones Universidad Alberto Hurtado, 2021.

Giesso, Osvaldo. *Recordando a Emilio Renart.* Espacio Giesso Reich, 1998.

Gómez-Sicre, José. *Emilio Renart of Argentina.* Pan American Union, 1965.

Lebenglik, Fabián. "Creatividad es la palabra clave: Homenaje a Emilio Renart (1921–91) en el Centro Borges." *Página 12,* February 19, 2002. https://www.pagina12.com.ar/diario/artes/11-2002-2002-02-19.html.

Pérez, Elba. "Cinco obras de Emilio Renart condenadas a la destrucción." *Tiempo Argentino,* March 23, 1985, 9.

Pratt, Mary Louise. *Planetary Longings.* Duke University Press, 2022.

Renart, Emilio. "Alunizar desde la Tierra." Interview by Carlos María Caron. *2001: Periodismo de Anticipación,* October 1969, 52–57.

Renart, Emilio. *Multimágenes.* Exhibition brochure. Galería Arte Nuevo, 1980.

Renart, Emilio. *Renart expone.* Exhibition brochure. Galería Pizarro, 1962.

Smith, Terry. "Defining Contemporaneity: Imagining Planetarity." *Nordic Journal of Aesthetics* 49–50 (2015): 156–174.

Spivak, Gayatri Chakravorty. *Death of a Discipline.* Columbia University Press, 2003.

Squirru, Rafael. "Pop Art or the Art of Things." *Americas,* July 1963, 15–21.

Vidal Mackinson, Sebastián. *Alienígena: Emilio Renart y su práctica artística y social.* Colección de Arte Fortabat, 2024. https:www.coleccionfortabat.org.ar.

Vidal Mackinson, Sebastián. *Emilio Renart: Integralismo, Bio-Cosmos, 1962–1967.* Museo de Arte Moderno de Buenos Aires, 2016.

Villa, Javier. *Constancia de la especie: El archivo de Emilio Renart.* Exhibition brochure. Galería del Infinito, 2024.

IV

Human and Nonhuman Histories

11

Exhibitions in the Face of the Climate Crisis

How Is the Anthropocene Shaping Hemispheric Aesthetics?

Valeria Meiller

The apocalyptic vision of an unlivable earth has long been crystallized in the public's imagination. More recently, it has increasingly assumed a dominant position in various institutions' agendas. In 2012, the *New Yorker* published a Tom Toro cartoon in which a man in a ragged suit sits with three kids, their clothes also in tatters, around a campfire in a dismal landscape. With a blurry skyline of a city in ruins, the caption reads, "Yes, the planet got destroyed. But for a beautiful moment in time, we created a lot of value for shareholders." Shared by climate activists ranging from Greta Thunberg to US Senator Bernie Sanders, the cartoon became a grim commentary on a crisis that demands an urgent shift of current economic and political paradigms yet is constantly being recaptured by the claws of capitalism. Everything, the cartoonist seemed to suggest, gets reabsorbed by the web of a system that keeps producing *more* ("value for the shareholders"), when it has long become clear that transforming the economy is among the only effective strategies to fight the climate crisis.

Recently I encountered a variation of Toro's original cartoon on social media platforms. It reads, "Yes, the planet got destroyed. But for a beautiful moment in time, we were able to respond to world problems with ecological exhibitions." This new version of the *New Yorker* cartoon signaled the potential—or lack thereof—of this moment in art history. During the 2020s we have seen environmental discourses emerge more explicitly as part of art institutions' agendas, curatorial discourses, and art practices. This engagement with the reality of climate change and the efforts to address the current environmental crisis from the materiality of artistic practices appear as a natural response. But is it? In this article I focus on contemporary art exhibitions that foreground environmental aesthetics from a hemispheric perspective. Specifically, I analyze art exhibitions from the 2020s that serve as examples of how art institutions are engaging with academic, political, and activist discourses on the environmental crisis. Of particular interest is how these practices put forth novel transnational efforts to address global North-South dynamics, the contribution of curatorial discourses to ecological conversations, and the limitations and potential setbacks of these institutional endeavors.

I enter a conversation pioneered by other Latin American scholars who have addressed the role of literature and the arts amid the current environmental crisis. Mariano Siskind suggests that literature and the arts have become sites where we mourn and grieve the end of the world:

> I would like to suggest the possibility of considering literature and the arts (and the very specific kind of attention we pay to them) as sites where we find ways to mourn the world, of the imaginary structure of an impossibly universal, emancipated community to come that we know is lost forever; mourning without closure, that is a melancholic kind of mourning which cannot withdraw the libido from the vanished object because losing the world (losing the very structure of political utopianism) is not the kind of loss that can be overcome. (226)

This sense of loss and mourning is, in Siskind's view, the primary form of agency of literary and artistic production in what he summarizes as "the experience of the end of the world" (227). Before moving forward, I would like to underscore the importance of collective responses such as mourning and grieving in the elaboration of environmental trauma. These approaches inform artistic practices in a broad sense, engaging in meaningful ways with coping, healing, and repairing historical and political trauma as experienced by many marginalized communities.[1]

In the specific case of environmentally induced trauma and the overall question about what artistic vocabularies can bring to the table, other scholars have signaled the importance of art as a site of visibility. In *Lo que el arte ve, lo que nosotros no vemos* (What Art Sees, What We Don't See, 2022), Graciela Speranza offers an example of this viewpoint. The title of her book already conveys that art might be able to visualize something that we humans are otherwise unable to perceive. In Speranza's view, contemporary literature and the arts are visualizing devices of the invisible; they can offer a glimpse into the true scope of the environmental crisis. At the core of Speranza's work is the question about matter and abstraction: How do we make visible in a material way what is ungraspable at a human scale? Though she is not the first to point this out, Speranza rightly underlines the value of visibility as a privileged political strategy of the present: "The political power of art lies, precisely, in reorganizing the realm of the sensible, modifying the visible, and the ways it is perceived and expressed. Aesthetics are a kind of interface. And more: Art cannot only reveal what we do not see and modify the sensible, but it can stare into the present and probe its darkness" [La potencia política del arte radica entonces, precisamente, en reorganizar el campo de lo sensible, modificar lo visible, las formas de percibirlo y expresarlo. La estética es una suerte de interfaz. Y más: el arte no solo puede dar a ver lo que no vemos y modificar lo sensible, sino que puede incluso fijar la mirada en el presente y sondear su oscuridad] (20).

It is worth noting that Speranza's particular understanding of "darkness," a metaphor of the unseen, is deeply informed by Timothy Morton's concept of hyperobjects as objects bigger than human scale that are partially invisible to human perception: "How to stare at those viscous, interobjective objects that elude time and the human scale in the twenty-first century?" [¿Cómo mantener la mirada en esos objetos viscosos, interobjetivos, que escapan al tiempo y la escala humana en el siglo XXI?] (21). This orients visibility toward dimension and the ways in which we might or might not be able to grapple with large-scale climate-change phenomena such as sustained droughts, transnational wildfires, and severe storms like hurricanes and cyclones, among many others. If we validate Speranza's take on visibility as a privileged form of political agency, the question about the role of art in a time of environmental crisis might stand a chance of doing something other than grieving. Of course, there is some nuance to this. The modes in which the temporality and planetary scope of the Anthropocene are shaping hemispheric aesthetics do not propose a causal relationship between our new geological era and the practices of art institutions. It is not as simple as proposing that climate change is creating a particular form of art or that contemporary art practices are crafted as responses that invalidate previous modes of aesthetics. In many ways, the new paradigm in art museum practices is deeply informed by modifications taking place outside of the art sphere. Particularly, it is related to a shift in the role of natural history, as society comes to the realization that the paradigm of the natural sciences alone is falling short in accounting for the social and historical scope of the environmental crisis. Informed by these shifts in both natural sciences and the arts, I divide this article into four sections to tackle individual exhibitions that overlap with vocabularies from other disciplines, spheres, and conversations to investigate their present roles and their responsibilities in this time of sustained ecological collapse.

In the first section I explore how contemporary curatorial narratives incorporate theoretical ecological frameworks of great complexity, often originating in academic circles, as a strategy to shed new light on art materials from the past and present. As an example, I tackle the exhibition *SimBiología: Prácticas artísticas en un planeta en emergencia,* held in Buenos Aires in 2021–2022. In the second section I analyze *El Dorado,* a series of exhibitions in the United States and Latin America that follow the hemispheric routes of gold from colonial times to the present. Informed by notions from material culture, I utilize an object-oriented framework of analysis to underscore how matter can convey histories of colonial and environmental violence while at the same time becoming the carrier of political resistance for marginalized and oppressed communities. Then, I turn to *no existe un mundo poshuracán,* a show that dealt with the aftermath of Hurricane Maria in Puerto Rico from a geopolitical perspective. By engaging with contemporary events at the intersection of art and political ecologies, this exhibition held at the New Museum in 2022–2023 interrogated the potentialities of contemporary Latin American and Latinx art in transnational and hemispheric contexts. Finally, in the last section, I engage with notions of communal resistance and vegetal resilience to explore the work and public program of *Life Between Buildings,* an exhibition at MoMA PS1 in 2022–2023 that focused on the political value of interstitial spaces in the lived environment.

(New) Conceptual Approaches

The languages of critical theory, philosophy, and art practice have always been intertwined. Yet, we might be living in a time where the explicit overlap of these languages is more tangible than ever. In the aftermath of the Covid-19 pandemic, the feminist waves of the previous decade across the Americas, the Black Lives Matter movement in the United States, and the rapid acceleration of the climate crisis on a planetary scale, institutional and curatorial endeavors are responding to social and political issues with a particular sense of urgency. In this context, interventions about the environmental crisis are proliferating in contemporary museums' agendas, expanding the scope of preservation as it has been traditionally understood in art history to engage in discussions about the preservation of species, ecosystems, and the future of our planet. These efforts are geared toward creating embodied critical frameworks to understand what the implication of "art preservation" might be and look like in the context of this new geological era. The extent to which these transformations have tokenized and reductively underlay agendas—related to these institutions' requirement to remain timely and politically relevant—needs to be more thoroughly discussed. So does the impact these institutional efforts may have in shifting the climate crisis conversation inside and outside the art sphere. Yet, some exhibitions are worth foregrounding as examples of how new critical vocabularies, often produced and circulated mainly within academic circles, are permeating the art world to inform new ways of understanding art materials, both present and past, in an ecological sense.

As an example, take *SimBiología: Prácticas artísticas en un planeta en emergencia,* a show curated by a team under the direction of Valeria González at the Centro Cultural Kirchner in Buenos Aires from October 2021 to June 2022. *SimBiología* comprised about 170 works by Argentine contemporary artists produced from the 1960s onward. The neologism *SimBiología* transformed the traditional signifier *simbología* (symbology) to include the meaning of *simbiosis* (symbiosis), thereby creating a semantic field where the evocative capaciousness of symbols could also unfold as relational, mutualistic relationships between species. The curatorial texts state that the show explored new links and symbiosis between the human and the nonhuman in a historical context that calls for reassessing our modes of thinking and being in the world. With this purpose in mind, the curators created a "Glosario" (Glossary) encompassing 56 environmental concepts. Spanning more general notions like "ambiente" (environment) to more specific ones like "actantes no humanos" (nonhuman actants), the glossary created conceptual umbrellas for the exhibited works. Under the category "Derechos de la naturaleza" (Nature's Rights), for instance, the exhibition's glossary referred to how countries such as Ecuador, in 2008, and Bolivia, in 2019, enshrined nature as a rights-bearing subject in their constitutions. This change underscored a shift from an extractive view of nature as a resource to a more nuanced way of thinking about the human and the nonhuman as interdependent.

These new constitutional frameworks reflected the legislative efforts of certain South American countries to acknowledge the rights of diverse forms of life and foregrounded the beliefs upheld by the Indigenous nations within those nation-states. This approach, which is key to tracing decolonial paths in Latin American environmental politics, also informed some collaborative projects that rethink the relationship between the human and the nonhuman, with an emphasis on territorial notions.

These works included the artistic practice of the duo composed of Guillermo Faivovich and Nicolás Goldberg. For almost two decades, Faivovich and Goldberg have been producing a body of artwork based on their research of *Campo del Cielo,* the meteorite field that lies in the provinces of Chaco and Santiago del Estero in Argentina. This art project began in 2006 as "Una Guía a Campo del Cielo" and so far encompasses materials as varied as the 2010 installation of two halves of a 1,574 kg meteorite, El Taco, at a gallery in Frankfurt; the publication of a book, *The Campo del Cielo Meteorite, Vol. l: El Taco,* also in 2010; and more recently, the exhibition *Otumpa* at the Museo Moderno de Buenos Aires in 2024. The project conveys an artistic attempt to historicize the institutional efforts and omissions in preserving the findings of *Campo del Cielo*. Given the new interest in geological narratives of the present, Goldberg and Faivovich's artwork serves as an example of how nonliving elements confront geopolitics at a planetary scale. Other projects from the *SimBiología* exhibition working along these lines include the socioenvironmental cartographies generated by the Iconoclasistas, a collaboration of Julia Risler and Pablo Ares. In 2009–2020 they generated a map library comprised of drawings, maps, photographs, and digital models as a collaborative research device of territorial activism. This group, modeled after the participatory research action of Orlando Fals Borda, aims to produce a dynamic perception of territories and collective meanings across disparate spheres such as neighborhoods, museums, academia, and activist networks.[2]

The intention to highlight a critical vocabulary around art practices was also enhanced by the delay in the exhibition opening date caused by the pandemic. This prompted the curatorial team to share their progress toward holding the show by publishing an archive of readings that might have otherwise remained as part of the internal research process. Under the title "Prácticas artísticas en un planeta en emergencia" (Artistic Practices on a Planet in Emergency)—with the Spanish word *emergencia* referring to both a sense of emergency and an emerging critical vocabulary related to the environmental crisis—the curatorial team made available a series of texts that united the voices of Latin American, European, and North American thinkers that reworked the vocabularies of humanist approaches to formulate posthuman and anthropodecentric epistemologies. This section reunited thinkers such as the Brazilian anthropologist Eduardo Viveiros de Castro, who put forth Amerindian cosmologies to rethink the scope of nature, and US feminist theorist Donna Haraway, who presents new modes of species relationships between human and nonhuman animals. In sum, it created an archive of conceptual contemporary approaches that aim to reimagine the vocabularies through which people have historically understood human relationships with the natural and cultural realms.

In an article in *Revista Ñ*, González states that the exhibition aimed to dismantle the idea that the environment preceded humans to "undo, through art, the disassociation between subject/object as the primary mold of every perception" [desandar, a través del arte, esa disociación sujeto/objeto como molde primario de toda percepción] (in Casanovas). The ambitious goal of shifting a viewer's perception aligns with contemporary theories of modernity that call for dismantling the division between nature and culture to instead propose the intertwining of these categories as indivisible (Latour). The complexity of this exhibition's pieces would deserve a chapter of its own here, but the focus of this analysis is to index the complex theoretical thread braided by the curatorial team to reframe the works exhibited under an environmental ethos. While many of the pieces exhibited in *SimBiología* might not have been originally conceived as "environmental art," the curatorial narrative created a strong case for why they could be revisited as such.

Furthermore, the digital archive hosted by the exhibition—gathering concepts and texts from diverse communities and geographies, many of which had not previously circulated in Spanish—created a novel narrative for Argentine art produced in the twentieth and twenty-first centuries. By rearranging a large body of artwork from the previous eighty or so years according to contemporary environmental vocabularies, *SimBiología* serves as an outstanding example of how Latin American art is engaging with ecological conversations of the present and fostering hemispheric philosophical dialogues.

(New) Material Approaches

Material and object-oriented viewpoints are also represented within this broader trend. New methodological approaches in scholarship since the 1980s have shifted attention to how matter informs understandings of culture and contest previous idealist propositions—what we have come to call the "material turn." Art institutions have also started to inquire about the capaciousness of matter to narrate historical and political processes across the hemisphere. Of particular relevance is the example of *El Dorado,* a hemispheric exploration of gold across art institutions of the Americas. Orchestrated as three independent exhibitions in Argentina, Mexico, and the United States in 2023–2024, this curatorial project embarked on the ambitious enterprise of understanding the conceptual and material scope of gold and how the myths surrounding this malleable metal can serve as a gateway to organize a particular narrative of the Americas.[3] From commodity to myth, object of desire to national reserve, gold serves as a concrete yet expansive category that threads histories spanning from pre-Columbian cosmologies to contemporary histories of extraction.

The multiplicity of vantage points that these exhibitions propose for gold can be understood as part of the ontological and epistemic shifts proposed by material culture, but also from decolonial methodologies. Consider Walter Mignolo's view on the importance of aesthetics in dismantling hegemonic narratives: "If colonial wounds are consequences of systemic and hierarchical social classifications, and social classifications are hierarchical epistemic inventions disguised as representations, then healing colonial wounds becomes a matter of epistemic and aesthetic reconstitution" (11). One of the greatest accomplishments of these exhibitions in this regard is that instead of exclusively being organized around histories of colonial dispossession, the works included also reflect on how the myth of El Dorado exists within the coordinates of neoliberal capitalism. As a framework of analysis that organizes the past and present of hemispheric dialogues, *El Dorado* foregrounds a manifold understanding of culture, nature, and the politics of life in the territories of the Americas.

The first of the exhibitions was held at Fundación Proa from April 1 to August 6, 2023. Fundación Proa is located on the waterfront of the Riachuelo, one of South America's most polluted rivers, in the neighborhood of La Boca in Buenos Aires. As the place where the Spanish landed in 1536, Proa's location stands as a testimony of colonial plundering as well as the tangible result of centuries-long devastating environmental policies.[4] Titled *El Dorado: Un territorio,* the exhibition posed the question of whether El Dorado existed within the territory of the Americas.

In an interesting twist, the show reframed the question about the importance of gold to the richness and vastness of America's resources. Despite not finding El Dorado as the myth proposed, colonial settlers did encounter wealth, seemingly infinite resources in the Americas. Adriana Rosenberg, director of Fundación Proa, explicitly enunciates the intention of rethinking the scope of the myth and argues that the extraction of gold alone does not account for the signifier of El Dorado. In her view, this mythological site can also be understood as a "fertile, exuberant, rich, and generous territory" [un territorio fértil, exuberante, rico, generoso]. Rosenberg questions what treasures El Dorado might represent in contemporary times: "What is El Dorado? What contemporary treasures does it refer to?" [¿Qué es El Dorado? ¿A qué tesoro remite en la actualidad?] (in Fundación Proa).

The reframing of El Dorado from a territory rich in gold to a territory rich in natural exuberance is exemplified by the materialities engaged in the works exhibited.

> Starting from what we call "ground zero" as equivalent to the territory, that element of the landscape that seemed deserted and contained so much wealth, we decided to focus the vision and reflection on the valuable substances of the continent—tomato, potato, cocoa, natural rubber, etc.—and display the diversity and exuberance of a geography rich in minerals, flora, and fauna that revolutionized the world order.

> Partiendo de lo que llamamos el "grado cero" como equivalente al territorio, ese elemento del paisaje que parecía desierto y que contenía tantas riquezas, decidimos concentrar la visión y la reflexión en las sus-

> tancias valiosas propias del continente—tomate, papa, cacao, caucho, etc.—, y mostrar la diversidad y exuberancia de una geografía rica en minerales, flora y fauna que revolucionó el orden mundial. (Rosenberg, in Fundación Proa)

Focusing on contemporary works, the exhibition featured materials as diverse as potatoes, corn, gold, and crafts as examples of the incalculable wealth that the Americas offered settlers and continue to offer the world. Mexican artist Betsabeé Romero, for example, works through the history of rubber extraction using car tires to create sculptures that critique the legacy of extraction for the automobile industry in Latin America. This issue is at the core of Latin American novels such as that of Colombian writer José Eustasio Rivera's *La vorágine* (1924), making Romero's work a link in a long tradition on the subject.

Nonetheless, the material conditions behind decolonial endeavors like the ones proposed by Proa's exhibition are anchored in the circulation of present-day symbolic and concrete power and capital relations. While this exhibition engaged with material questions, it also complicated the criticisms it was meant to highlight. As part of the ever-evolving paradoxes behind art institutions, in her coverage of the show in *e-flux*, Sylvie Fortin writes,

> Fundación Proa is a private entity essentially supported by Tenaris and Ternium, two divisions of Techint, a building and engineering multinational whose activities, anchored in steel mining and production in Latin America, encompass manufacturing, oil and gas, health care, and technology, media and environmental remediation ventures. Ternuim's and Tenari's logos grace the exhibition title and are twice featured in the catalog. And Fundación Proa shines on the Community Relations section of Techint's website, where the company boasts its forty million US dollar investment in education and culture, reaching over 400,000 individuals. If gold was El Dorado's mirage, our social imaginaries are its new horizon.

Fortin's critique of the mirages of El Dorado as an example of the contradictions at play in present-day exhibitions also preoccupies theorists like Mignolo in his analysis of decolonial methodologies. A true shift in episteme requires, in Mignolo's view, not only a change in the contents of our inquiries but also a shift in the modes by which art is funded and circulated. The current endeavors of art institutions funded by corporate capitalism might not be radical enough. Those fundamental changes are key to producing a true epistemic reconstitution since decoloniality "is *about the terms of the conversation, not just the contents*" (Mignolo 13). The question is not only whether art institutions can be the sites that feature artistic practices that align with decolonial discourse. It is also whether the material conditions that allow these institutions to function are not themselves part of the problem.[5] This is, of course, a long-standing question that precedes and will outlive the particular case of Proa and for which in this article I do not pretend to offer a satisfactory answer.

The second iteration of El Dorado was a two-part group exhibition at the Americas Society that took place from September 2023 to May 2024.[6] Unlike Proa's exhibition, which provocatively reframed the idea of El Dorado as a promised land of exuberance and richness, this New York–based institution decided to take a conceptual route to engage with gold. Titled *El Dorado: Myths of Gold,* this exhibition focused on a historical revision of the myth of El Dorado to investigate how artworks "challenge, reinforce and question the continuity and effects of the myth in the Americas" (Americas Society). Perhaps the most striking aspect of this exhibition is how it updated the colonial enterprise's quest for gold to the current sociopolitical and environmental context. In this iteration of El Dorado, curators Aimé Iglesias Lukin, Tie Jojima, and Edward J. Sullivan decided to place historical works *vis-à-vis* contemporary works to shed light on the ways the myth has been updated, questioned, and contested through hemispheric art production. Particularly striking was the exhibition's engagement with cosmological orders, showing how Indigenous and colonial iconographies have resorted to gold as a medium to narrativize their worldviews. In their prologue to the catalog, Lukin, Jojima, and Sullivan recapitulate the most famous legend of El Dorado, which originated within the territory that coincides with present-day Colombia. The curators describe this myth known as the Golden King as "a ceremony of power investiture among the Muisca people whereby a new leader was covered with gold dust and placed on a raft adorned with treasures that would then be submerged, along with many gold offerings, in Lake Guatavita, near Bogotá" (9).

The tale, which was believed to have been confirmed by Spanish conquistadors, is the source of many pieces in the exhibition, including a small devotional object representing a raft cast in gold from an unknown artist circa 1600 CE and video documentation of the 1981 performance *Nubes para Colombia* by Pedro Terán, an artist from the Venezuelan conceptual art movement. Four centuries after the unknown artist created the gold raft, Terán goes back to the imagery of the Golden King, but instead of offering a triumphant image such as the one evoked by the Muisca leader navigating Lake Guatavita, he covers himself in gold dust as a commentary on twenty-first-century extractivism in

Latin America. Interpellating the condensed information of symbols, the artist began his performance dressed in the colors of the Venezuelan flag, then proceeded to undress himself, tie his genitals, and cover himself in golden dust. The statement of his piece was straightforward: it offers a commentary on the sacrificial nature of capital, which has defined the fate of Latin American nations such as Venezuela until the present day. But Terán's piece is striking because, despite conveying a harsh critique, it also becomes a luscious image of gold's beauty both as symbol and materiality. Conveying two very disparate pieces from very different time periods, this counterpoint effectively indexed the broad affordance of gold.

Something quite interesting also happed in the Americas Society exhibition's section devoted to Catholic imagery, where virgins and saints carved out of gold over diverse media configured another broad constellation of meanings. There, the audience could encounter an eighteenth-century canvas of a classical fair-skinned *Our Lady of Antigua*, to whom Spanish *conquistadores* were devoted, as she was known for offering protection against enemies, together with a Black *Our Lady of Regla,* modeled after a Yoruba goddess by Harmonia Rosales in 2019. Amid many other figures spanning from the sixteenth century to the present, this section of the exhibition spoke volumes about the long-standing relationships between gold, coloniality, and the Catholic imagination. But it also allowed the spectator to trace counterpoints, gender, and racial critiques within it by operating simultaneously at various timescales.

As a conclusion to this section, I propose to interpret *El Dorado* as an important linchpin to stimulate hemispheric dialogues and histories that continue to be traversed by questions about wealth. Despite the potential problems raised by the funding entities behind these exhibitions, *El Dorado* stands as a valuable effort to conceptualize and complicate the historical scope of long-standing historical narratives through concrete approaches to materiality. In recent years, academic interventions such as Héctor Hoyos's 2019 book *Things with a History* have foregrounded the importance of the material circulation of commodities in the crafting of symbolic cultural production in Latin America. Hoyos exemplifies this through literary texts, but his arguments—some of which engage specifically with gold—can also be extended to other disciplines like art production. These efforts, which we see materialized in *El Dorado,* explicitly contest the bounded existence of national narratives to inquire about the transnational scope, hemispheric and global, of labor, commodities, and extraction. It also signals the malleability of these materials to inform but also resist and reevaluate what these narratives might look like in an age of ongoing environmental collapse.

(Urgent) Environmental Approaches

Some exhibitions from recent years have also been organized around contemporary environmental catastrophes, serving as urgent reflections on ecological disasters as well as tributes of remembrance for the human and environmental toll of these events.[7] From November 2022 to April 2023, the Whitney Museum held *no existe un mundo poshuracán: Puerto Rican Art in the Wake of Hurricane Maria,* a group exhibition reuniting intergenerational artists from Puerto Rico and the Puerto Rican diaspora on the fifth anniversary of Hurricane Maria, the deadly category 5 hurricane that devastated the Caribbean and Puerto Rico in particular in September 2017. The exhibition reunited works produced in the span of the five years before the opening and, according to the curatorial text accompanying the show, was the first exhibition held at a large US museum focused on Puerto Rican art in nearly a half century (Whitney, "no existe"). Besides the explicit intent of memorializing and acknowledging the devastating impact of Maria, the exhibition's pieces straightforwardly denounced the political, cultural, and economic neglect by the United States toward Puerto Rico, an unincorporated US territory with commonwealth status. The disenfranchisement of Puerto Rico in US politics as a nonvoting territory with no representation in the US Congress is part of a long conversation about the forces of neocolonial power. Acquired by the United States from Spain after the 1898 Spanish-American War, the territory has long suffered from economic depression, a shrinking population due to migration, government mismanagement, and lack of economic sovereignty. These issues intensified with the Covid-19 pandemic and the ongoing effects of natural disasters.

The close yet fraught relations between the United States and the island is at the heart of the work of many Puerto Rican artists and writers who have long criticized the occupation. An example is Puerto Rican writer Giannina Braschi's 2011 novel *United States of Banana,* whose allegorical plot follows the character of Giannina, an alter ego of the author, as she attempts to free Puerto Rico from its colonial captivity. US occupation is also a central theme for Roque Salas Rivera, a Puerto Rican poet whose highly political work focuses on the experiences of colonialism and migration of the Puerto Rican diaspora, subjects that are expansively explored in his volumes *x/ex/exis: poemas para la nación* (2021) and *antes que isla es volcán* (2022). Salas Rivera's critical stance on the conflict is not only a theme within his work but also a defining element in the bilingual power dynamics that shape the formal aspects of his poems. In his 2019 collection *While They Sleep,* Salas Rivera writes most poems in English and uses Spanish footnotes in a much smaller font size. There, colloquial Puerto

Rican Spanish functions as a commentary and counterdiscourse to the information provided by the English text. Take the untitled poem where he writes, "one week after the hurricane we / have different views on colonialism" and in footnote 28, "As if you cared about my people's name" [Como si te importara el nombre de mi pueblo] *(While They Sleep* 48). Here, the polarization of perspectives in the aftermath of the hurricane is quickly undermined by a note that underscores how Puerto Ricans' ("mi pueblo") take on the US occupation goes largely unnoticed by official historiographies.

Because of the profound impact of Salas Rivera's experimental poems, the Whitney Museum decided to use a line from his work to name the exhibition ("no existe un mundo sin huracán"). This was a gesture toward carving out a space to inscribe the alternative, silenced vision of the colonized.[8] Matters can, of course, be complicated. The question about the subaltern is not necessarily the history of a happy restitution. While the gesture of redemarcating the confines of the American museum space represents the unrepresented, Salas Rivera's line signals the unworlding of the world. For Salas Rivera, there is no world in the aftermath of the hurricane. What does this American museum, standing as an evocative ship facing the Hudson River, suggest about the work exhibited? Do these works do anything other than memorialize catastrophe and signal, once again, a crisis of futurity? To a certain extent, here the odds seem to be in Siskind's favor.

(Non/Human) Histories at the Museum

While some contemporary exhibitions focus on large-scale environmental disasters, others shed light on the ecological small-scale contributions of nonhuman elements to the regulation of lived environments. Take the example of *Life Between Buildings*, an exhibition at MoMA PS1 in New York City running from June 2022 to January 2023. Organized around artists' engagement with interstitial urban areas such as "vacant lots, sidewalk cracks, traffic islands, and parks," the exhibition focused on how artistic intervention can turn these "negative spaces into sites for common life: gardens, installations, performances, and gatherings" (*MoMA PS1*, "Life Between Buildings"). Community gardens reemerged as part of the city's overall efforts to cope with the Covid-19 pandemic, which included, among its many devastating social effects, a spike in food insecurity and a ban on indoor group gatherings.[9] Inspired by this renewed interest in community gardens, *Life Between Buildings* gathered art produced in New York City from the 1970s onward that stepped outside of studios, galleries, and museums to reimagine urban spaces through "community efforts to rethink the cityscape, and recovering space towards creative, communal, and ecological ends" (*MoMA PS1*, "Life Between Buildings").

Despite the historical perspective offered by this exhibition, the explicit demarcation of its critical efforts as part of contemporary vocabularies that rethink human/nonhuman entanglements in the art sphere was clear in the ecological lens assumed by Jody Graf, the assistant curator at MoMA PS1 who oversaw executing the show, but also by the partnerships that financed its execution. In this case, the funding partner was Allianz, a multinational insurance company based in Germany that has a public record of financing ecologically driven institutions as part of an overarching commitment to become a "steward of nature" (Airmic) and a commitment "to limiting global warming and fostering a just transition to a low-carbon future" (Allianz).

Central to this exhibition's inquiry was the role of community gardens as spaces of collective resistance in urban settings. It focused on what the resilience of the vegetable world and grassroots movements might be able to teach about ecological resistance.[10] *Growing Abolition* was an installation at the MoMA PS1 courtyard made in collaboration between artist jackie sumell and the Lower East Side Girls Club community center. A reflection on abolitionist ideas through space and the vegetal world, the artist created a "solitary greenhouse" upon the blueprint of a solitary confinement cell in ADX Florence, an administrative maximum federal prison near Florence, Colorado. This prison, unofficially known as "the Alcatraz of the Rockies," is a maximum-security complex for male prisoners that has been involved in public controversies for many years. The reasons for this facility's exposure include legal suits alleging the prison's failure to provide proper mental health care to seriously ill inmates and chronically abusing prisoners by means including prolonged lockup in solitary confinement in underground cells that torture prisoners by depriving them of exposure from daylight cycles. This is the backbone of *Growing Abolition*, which is the product of a seven-year-long collaboration of sumell and an inmate called Herman Wallace, who was kept in solitary confinement for 27 years at ADX Florence in 6 × 9 foot cells for a minimum of 21 hours a day during his incarceration for 41 years of his life. In this solitary greenhouse, plants become beacons of survival that signal the possibility of reconvening and repopulating torture spaces with growing plants, sunlight, and overall, the promise of life. In a conversation with Mina Stone, the owner and chef at PS1's restaurant, sumell expands the scope of her collaboration to include plant and garden suppliers as active contributors to the piece and manifests her desire that the larger society can continue to "seed abolitionist ideas and continue to grow them" (in MoMA PS1, "Growing Abolition").[11]

The agency of vegetal life as a means of political resistance is also at the core of other projects featured at the exhibition. One of the interviews included in the digital portion of the show featured Peter Cramer and Jack Waters, who since 1995 have been running the public community garden *Le Petit Versailles* on East Second Street as a space devoted to resisting social inequities in the Lower East Side. As founding members of the performance group POOL, Cramer and Waters staged performances broaching issues related to class, race, sexuality, and urban ecology in this public garden. The lot where *Le Petit Versailles* stands—originally squatted in 1995, the same year that HIV retroviral drugs began being disseminated—became a survival space for New York City's gay community during the AIDS epidemic. The occupiers' intentions were to recuperate a public space while also creating a sense of unity, intimacy, and safety among the queer community, intentions that regained significance and momentum during the Covid-19 crisis.

Public programs were also at the forefront of *Life Between Buildings*, with interventions engaging a wide variety of events that included curatorial conversations, poetry readings, and performances. In the vein of expanding art exhibits toward public and community-engaged work, these interventions reframed the scope of the materials exhibited inside the museum by showing their entanglements with larger structures within the cityscape. "The Art of the Land" was a two-panel discussion organized by PS1 and the Clemente Soto Vélez Cultural and Educational Center, a Puerto Rican/Latinx multiarts cultural institution in Loisaida,[12] the Lower East Side, that has been dedicated to broadening the cultural vision of the city's artscape and has served as a platform for New York-based Latinx and BIPOC artists over three decades.

This two-panel intervention broadened the understanding of "interstitial spaces" that the works in the museum investigated. In the first section, "Institutions beyond Institutions: Reimagining Architectures of Bureaucracy," the conversation gravitated around how, from the 1970s onward, "municipal and city-owned buildings across New York had been reclaimed by artists and curatorial workers as sites for experimentation" (MoMA PS1, "The Art of Land"). Besides the PS1 and the Clemente, which continue to operate in the buildings of decommissioned public schools, the explored case studies included El Bohio and ABC No Rio. The former had been an operating community center in the East Village from 1906 until 1977, when it fell into disrepair and became a drug house. In 1979, local residents, many of whom were Latinx immigrants, reclaimed it as a cultural center. From 1979 until the enforced eviction of its occupants in 2011, El Bohio functioned as a cultural center that flourished as a communal organization.[13] The interstitial nature of this space was already conveyed by its name. "El bohio," which could be translated in English as "the hut," signaled its existence as a safe space that gave continuity to the cultural legacy of the Latinx community of Lower Manhattan.

ABC No Rio is a collectively run nonprofit art organization that remains operational in New York City's East Side. From 1983 to 1991, ABC No Rio was run by Cramer and Waters, who in 1995 also founded *Le Petit Versailles*. This overlap serves as an example of how reclaiming public space played a fundamental role, spanning across practices as diverse as squatting a building and reclaiming a parking lot. ABC No Rio's first event was an exhibition entitled *The Real State Show*, a squatter exhibition organized by the artists group Collab, formed in 1977 and held at 156 Rivington Street. The exhibition tackled landlord speculation in real estate, which had already become an issue of concern in the 1970s, raising awareness about the power disparities at play in housing and ongoing gentrification. When the exhibition was evicted, the space continued to operate as an art gallery, resisting legal disputes with the city government until the city sold the building to ABC No Rio in 2006, after a long process that had originally begun in 1977. Today, an ambitious renovation project is being designed by architect Paul Castrucci. The new building, which will replace the distressed and disrepaired grounds of 156 Rivington, is based on energy-efficiency principles such as low energy consumption, the use of local and recycled materials, and a commitment to installing a green roof. The environmental dimension of this renovation continues to underscore ABC No Rio's original commitment to political ecologies in urban planning.

Life Between Buildings explores themes and materialities at the intersection of social and environmental justice. While other shows such as *no existe un mundo poshuracán* engage with large-scale environmental struggles, this exhibition looks at local modes of resistance carried out by different marginalized communities. Including examples from the LGBTQ+ community, grassroots movements, and contemporary abolitionist activists, the works and ideas portrayed at PS1 were deeply informed by the power of solidarity, resistance, and resilience. Focused particularly on the use of public spaces, with gardens as one of the core spaces featured, the works exhibited in *Life Between Buildings* signaled the importance of interstitial spaces—often disregarded in conversation about the large scale of environmental disaster—in the fight for climate and social justice. In doing so, it becomes an exhibition that might have something to teach about how artistic practice can remove the veil covering what we humans cannot see.

Beyond Visualizing Loss

The scope of contemporary exhibitions engaging with the current environmental crisis is vaster than this chapter's examples account for. Regardless, these four examples allow me to offer some conclusions to my initial question on how the Anthropocene might be shaping hemispheric aesthetics. First, I would like to suggest that current curatorial and institutional endeavors on environmental issues not only function as sites of mourning and visibility but play an explicit role in reimagining the narratives through which we look at art from the past and present. The political momentum behind these exhibitions seeks to intentionally partake in environmental politics beyond the confines of the art sphere. These shows dismantle any illusion of autonomy, setting the tone for a conversation that will likely continue to become more politically driven in years to come. Consequently, we might, in turn, begin to think differently about the relationship of politics and aesthetics and be prompted to reimagine the ontological limits of art objects as something other than art: temporal devices, narratological objects, metonymic expressions of the greater than human forces that tell the story of ecological collapse. Especially in the cases where institutions engage with the creation of catalogs and conferences—what we could call "discursive archives"—it becomes clear that efforts are geared toward rearranging cultural narratives in the face of the new climate regime.

In certain ways, the exhibitions analyzed here can be seen as a response to the predicament described by Amitav Ghosh in *The Great Derangement* (2016), in which he asserts that the extreme nature of today's climate crisis challenges the limits of the contemporary imagination. The Anthropocene, Ghosh writes, "presents a challenge not only to the arts and humanities, but also to our commonsense understandings and beyond that to contemporary culture in general" (9). Ghosh was among the first writers to underscore the importance of the arts and literature in understanding climate change. Besides observing that literature tackling climate change, mainly science fiction, had historically not been regarded as serious literature, Ghosh makes a claim for the importance of present cultural endeavors to question the intricate modes by which "the climate crisis is also a crisis of culture" (9). He rightly argues that literature and the arts play a fundamental role in shaping people's affects and desires, conjuring art's intimate entanglements with wider histories, specifically those of extractive systems fostered by imperial and capitalist regimes. In that vein, these exhibitions' value might reside in becoming aiders of new forms of imagination that can contribute to recognizing the ecological dimensions of the dire reality at hand and at least symbolically rectifying the omissions of hegemonic and anthropocentric histories.

Second, I would like to suggest that these new inclinations to rewrite art history are not and never will be completely independent from the power dynamics that have created the environmental challenges we face. Circling back to Toro's cartoon in this chapter's opening vignette, we need to acknowledge that while we can be enlightened by the complex conversations that art exhibitions are able to bring forth, we are also obliged to recognize—as is palpable in the examples of Group Techint funding Proa's *El Dorado* or Allianz's support to *Life Between Buildings* as part of its broader environmental campaigns—that these exhibitions are often made possible by corporate funding, confronting us with the thought of how art institutions are partaking in the current social and economic model instead of fighting it. The way in which these dynamics complicate the morality of their ecological discourse is not my main goal in this chapter. Nor is it the desire to cancel or dismiss these exhibitions' monumental efforts and beneficial outcomes. It just serves as an opportunity to continue to untangle the broader question, not to be answered here, about independent practices and the overall autonomy of art as a discipline. These economic dynamics behind the exhibitions I analyzed are part of the status quo of public and private institutions and the coordinates of the overall economy. As such, they are worth mentioning as we try to find a provisional answer to the question of whether humans can continue to respond to the environmental crisis with more art exhibitions.

Finally, I would like to pinpoint that these exhibitions stand as archival devices of a particular moment in time that I suspect we will turn back to as we explore the perception of humans, nonhumans, and the earth from some future time. Beyond their capacity or not to do anything in particular—if we can continue to demand that art "do" things for humans—these exhibitions' traces will exist as historical and ecological reservoirs of the images, languages, and ideas with which we made sense of this moment of the ongoing end of the world.

Notes

1 As an example, take the group exhibition *Grief and Grievance: Art and Mourning in America*, curated by Okwiu Enwerzork at the New Museum from February 17 to June 6, 2021. Bringing together thirty-seven artists working in a variety of media, this intergenerational show focused on the concepts of "mourning, commemoration, and loss as a direct response to the national emergency of racist violence experienced by Black communities across America." Particularly interesting was the scope of the affects elicited by the exhibition, which was set to work around representations that showed how "mourning can be seen as a distinct form of politics, one that refuses melancholia in favor of multifaceted forms of critique, resistance, and care" (New Museum).

2 Orlando Fals Borda was a Colombian sociologist who developed a sociological "participatory research-action" model. This model entailed not only research into the needs of communities but the development of action models to improve these communities' lives. Considered a pioneer of the decolonial research model, Fals Borda's work with campesinos in rural Colombia has become an inspiration for many subsequent researchers. For a comprehensive understanding of Fals Borda's work, see Rappaport.

3 This project was spearheaded by an academic symposium organized by Edward J. Sullivan, a professor of art history at New York University, held during fall 2021 and spring 2022. As a co-curator of one of the three projected exhibitions, the two-part show *El Dorado: Myths of Gold* at the Americas Society (2023–2024), Sullivan's engagement with the ideation of this project also signals the theoretical scope of contemporary institutional efforts in addressing the social and political impact of curatorial narratives in an age of environmental collapse.

4 For a comprehensive account of pollution of the Riachuelo see "Contaminación del Matanza Riachuelo," at ACUMAR (Autoridad de Cuenca Matanza Riachuelo, the official site of the government of Buenos Aires), https://www.acumar.gob.ar/contaminacion-del-matanza-riachuelo/.

5 For an in-depth critique of the entanglements between corporate power and art institutions, it is worthwhile referencing the 2022 documentary *All the Beauty and the Bloodshed (Poitras),* in which the artist Nan Goldin foregrounds the relationship of world-renowned art institutions with the Sackler family, the pharmaceutical dynasty behind the opioid epidemic in the United States. As part of the actions and mobilizations carried out by Goldin with a large group of activists, the film features the public performative actions they carried out in museums like the Metropolitan Museum of Art in New York City, where protesters demanded that the Sackler family name be removed. The Sackler family were big donors to worldwide art institutions and often financed entire pavilions. Removing their name was seen as a first step toward dismantling these companies' corporate power and their influence within the art world.

6 A third exhibition on El Dorado was to take place in 2024 at Museo Amparo, in Puebla, Mexico. But as I finalized this piece in August 2024, the show had not yet opened and is therefore left out of the analysis.

7 The memorializing aspects of these recent artistic interventions are varied in scope. In 2021 the Museo del Barrio held its first triennial, *Estamos bien,* devoted to emerging Latinx and Latin American artists from the New York metropolitan area. Founded in 1969, the Museo del Barrio stands as the leading Latinx cultural institution in the United States devoted to the preservation of the cultural legacy of Puerto Rico and the Caribbean. As part of its closing events, the biennial organized the vigil "Los que mueren por la vida, no pueden llamarse muertos" (Those Who Died for Life Can't Be Called Dead) that honored "fallen environmentalists across the world, offering a space to build historical environmental memory and prompting healing justice for those on the frontlines" (Museo del Barrio). Held in front of an altar of colorful votive candles and natural offerings outside of the museum, participants were invited to leave offerings in the form of soil, water, seeds, flowers, minerals, stones, and other natural elements as gestures of reciprocity with the Earth and of solidarity with envi-

ronmental activists. The offerings were then used to nurture the soil for a memorial tree planted in East Harlem.

8 Poetry was omnipresent in the exhibition in more than one way. As part of the public program of *no existe un mundo poshuracán,* the Whitney Museum invited Roque Salas Rivera to curate a reading featuring Puerto Rican poets. "El bello no ser de nuestros cuerpos / Our Bodies' Beautiful Not Being: A Reading" (December 4, 2022) was inspired by lines from Julia de Burgos's poem "Nada (Nothing)": "If from the not being we come, and to the not being we march, nothing between nothing and nothing, zero between zero and zero and if between nothing and nothing nothing can exist, let's toast the beautiful non-being of our bodies" [Si del no ser venimos y hacia el no ser marchamos, nada entre nada y nada, cero entre cero y cero, y si entre nada y nada no puede existir nada, brindemos por el bello no ser de nuestros cuerpos]. Understood as a gesture toward the possibility of creating from nothingness, "making life, a way forward, an even beauty, despite colonialism's erasure and violence" (Whitney Museum), the vocabulary around the event continued to signal a crisis of futurity. While poets filled the void for more than an hour and a half of bilingual Spanish and English poetry readings, their work somehow continued to underscore a crisis of meaning.

9 For an example of how artistic expression accounted for the social, environmental, and political role of New York's community gardens during the pandemic, see *Your Mouth Has Power,* a collaborative film by food magazine *MOLD* and chef DeVonn Francis, founder of Yardy World, featured as part of the *Critical Cooking Show* at the 5th Istanbul Design Biennial (*MOLD* and Yardy World).

10 "What is the revolutionary potential of urban gardens? What can plants teach us about gentrification and community resistance?" were the guiding questions of one of the curatorial conversations, "Gardens and the Politics of Urban Space," held in June 2022 as part of the exhibition's activities (MoMA PS1). During the colloquium, curator Jody Graf conversed with artists Danielle De Jesus, Matthew Schrader, and jackie sumell. All engaged with public space through the vantage point of political ecologies, underscoring the importance of communal and green spaces as key interventions of communal resistance.

11 To learn more about the project, visit the Solitary Gardens website, https://solitarygardens.org.

12 The Lower East Side is also known as "Loisada," a term coined by Hispanic residents and activists from the Latinx grassroots movement of New York City.

13 For a comprehensive account of Bohio, see Safiyah Riddle's article in *The City.*

Works Cited

Airmic. "Moving ESG into the Mainstream: The Case of Allianz and Its Community." *White paper,* 2021. https://www.airmic.com/system/files/technical-documents/Airmic-ESG-Insights-white-paper.pdf.

Allianz. "Tackling Climate Change." N.d. https://www.allianz.com/en/sustainability/climate-change.html.

Americas Society. *El Dorado: Myths of Gold.* Exhibition description, 2023. https://www.as-coa.org/exhibitions/el-dorado-myths-gold.

Artishock: Revista de Arte Contemporáneo. "*SimBiología: Prácticas artísticas de un planeta en emergencia.*" 30 December 2021. https://artishockrevista.com/2021/12/30/simbiologia-practicas-artisticas-en-un-planeta-en-emergencia.

Braschi, Giannina. *United States of Banana.* Amazon Crossing, 2011.

Burgos, Julia de. "Nada (Nothing)." In *Song of the Simple Truth: The Complete Poems of Julia de Burgos.* Translated by Jack Agueros, 28–29. Curbstone Press, 2008.

Casanovas, Laura. "Artistas ante un planeta en peligro." *Revista Ñ,* September 29, 2021, updated May 9, 2022. https://www.clarin.com/revista-n/agenda/simbiologia-cck_0_ZwA2n-621.html.

Faivovich, Guillermo, and Nicolas Goldberg. *The Campo del Cielo Meteorite, Vol. l: El Taco.* Documenta (13)/Hatje Cantz, 2010.

Fortin, Sylvie. "El dorado. Un territorio." *e-flux, Criticism,* 2 June 2023. https://www.e-flux.com/criticism/543792/el-dorado-un-territorio.

Fundación Proa. "Palabras del comité curatorial." In "El Dorado. Un territorio." Exhibition page, 2023. https://proa.org/esp/exhibicion-proa-el-dorado-2-textos.php.

Ghosh, Amitav. *The Great Derangement: Climate Change and the Unthinkable.* University of Chicago Press, 2016.

Hoyos, Héctor. *Things with a History: Transcultural Materialism and the Literatures of Extraction in Contemporary Latin America. Columbia University Press,* 2019.

Latour, Bruno. *We Have Never Been Modern.* Translated by Catherine Porter, Harvard University Press, 1993.

Lukin, Aimé Iglesias, Tie Jojima, and Edward J. Sullivan, editors. *El Dorado: Myths of Gold.* Exhibition catalog. Americas Society, 2024.

Lukin, Aimé Iglesias, Tie Jojima, and Edward J. Sullivan. Prologue to *El Dorado: A Reader,* edited by Aimé Iglesias Lukin, Tie Jojima, Edward J. Sullivan, and Karen Marta. Americas Society, 2024, 7–17.

Mignolo, Walter D. *The Politics of Decolonial Investigations.* Duke University Press, 2021.

MOLD and Yardy World. *Your Mouth Has Power.* Film. *Critical Cooking Show,* 5th Istanbul Design Biennial, 2020. https://empathyrevisited.iksv.org/en/critical-cooking-show/26-your-mouth-has-power.

MoMA PS1. "The Art of Land: Co-Organized by MoMA PS1 and the Clemente." Series page. https://www.momaps1.org/events/185-the-art-of-land.

MoMA PS1. "Growing Abolition: jackie sumell and the Lower Eastside Girls Club." Installation page. https://www.momaps1.org/programs/11-growing-abolition.

MoMA PS1. "Life Between Buildings." Exhibition page. https://www.momaps1.org/programs/1-life-between-buildings.

Museo del Barrio. "Estamos bien: La Trienal 20/21." Exhibition page. https://www.elmuseo.org/exhibition/estamos-bien-la-trienal-20-21/.

New Museum. "Grief and Grievance: Art and Mourning in America." Exhibition page. https://www.newmuseum.org/exhibitions/view/grief-and-grievance-art-and-mourning-in-america-1.

Poitras, Laura, dir. *All the Beauty and the Bloodshed.* Film. Neon, 2022.

Rappaport, Joanne. *Cowards Don't Make History: Orlando Fals Borda and the Origins of Participatory Action Research.* Duke University Press, 2020.

Riddle, Safiyah. "In Epic Battle for CHARAS/El Bohio Building, Owner Gregg Singer Buys Time with Bankruptcy Filing." *The City,* March 22, 2023, https://www.thecity.nyc/2023/3/22/23652370/charas-elbohio-gregg-singer-bankruptcy-foreclosure.

Rivera Salas, Roque. *antes que isla es volcán.* Beacon, 2022.

Rivera Salas, Roque. *While They Sleep (Is Another Country).* Birds, 2019.

Rivera Salas, Roque. *x/ex/exis: poemas para la nación.* University of Arizona Press, 2021.

Siskind, Mariano. "Towards a Cosmopolitanism of Loss: An Essay about the End of the World." In *World Literature, Cosmopolitanism, Globality: Beyond, Against, Post, Otherwise,* edited by Gesine Müller and Mariano Siskind, 205–235. De Gruyter, 2019.

Speranza, Graciela. *Lo que no vemos, lo que el arte ve.* Anagrama, 2022.

Whitney Museum of American Art. "El bello no ser de nuestros cuerpos / Our Bodies' Beautiful Not Being: A Reading." December 4, 2022. https://whitney.org/events/el-bello-no-ser-de-nuestros-cuerpos.

Whitney Museum of American Art. "No existe un mundo poshuracán: Puerto Rican Art in the Wake of Hurricane María." Exhibition page, 2022. https://whitney.org/exhibitions/no-existe.

12

The Act of Collecting in Latin American Documentary

Antonio Gómez

The finding of a complete copy of Fritz Lang's 1927 film *Metropolis* in Buenos Aires in 2008 was celebrated worldwide as a priceless discovery, conducive to the restoration of a masterpiece of silent-era filmmaking to its original form.[1] Conceived by Lang as a 153-minute film, *Metropolis* was cut to 116 minutes shortly after its release to facilitate international distribution. This shorter version, later copied and transferred to multiple media, was shown throughout the twentieth century and became a vital piece of the cinematic canon. Until the Buenos Aires print was found by Fernando Martín Peña in the Museo del Cine, audiences had only seen the shorter, incomplete version of *Metropolis*. The Buenos Aires copy contains approximately twenty-five extra minutes that at the time of its discovery had not been seen in almost a century and are now part of the 148-minute *Complete Metropolis* released on DVD and Blu-ray by Kino Lorber in 2010.

While this could be considered a heartwarming tale of poetic justice and altruistic cinephilia, it is a bitter reminder of the asymmetries between metropolitan and peripheral archives. Peña had been aware of the existence of the print since the 1980s but had not been granted access to the collection where it could have been housed. It took a death, a donation, and the informal initiative of the museum's new director (Peña's ex-partner, Paula Felix-Didier) to locate the film and verify its uniqueness. Even though the print was a 16 mm transfer of a lost 35 mm positive, poorly done in the 1960s or early 1970s and in precarious condition, it is the only surviving copy of the extra footage. Excited and convinced that the news would be eagerly received, Peña immediately reached out to the F. W. Murnau Foundation in Germany, but they expressed no interest. It took Peña's persistence to capture the attention of the German institute. He showed a DVD copy to a colleague in Madrid, who then called a contact in Germany, and they finally acknowledged the finding. Peña understands their skepticism: "*Metropolis* is one of the most sought-after films in the history of cinema. They must have been contacted by many people over the years. I attribute it to that and to the thought that they must have had, 'Imagine if *Metropolis* were to be in Argentina'" [*Metrópolis* es una de las películas más buscadas de la historia del cine, debe haber habido mucha gente a lo largo de los años, se lo atribuyo a eso y a que deben haber pensado: "Mirá si *Metrópolis* va a estar en Argentina"] (in Barrera).

This anecdote highlights both the precarious state of Argentine museums and archives and the position that Latin America continues to hold in the global distribution of symbolic capital. It would have only taken basic cataloging to confirm the existence of this copy; instead, its verification required not only personal initiative but also rule-bending informality. The print was never truly lost; it was merely tucked away in an unsearchable collection. Once it was located, questions arose about its authenticity, and its value was downplayed. Again, an informal and creative approach was necessary to capture the attention of an institution with the resources to properly deal with such a finding. Restoring the film in Argentina was never considered an option, unlike what would have happened had the copy been in the United States, Japan, Australia, or any other developed economy. In Peña's account, the agreement between the Museo del Cine and the Murnau Foundation, which legally owned it at that moment, was very beneficial for the Argentine institution, as it allowed for the preservation of 9,000 meters of Argentine material that was at risk (Peña 84).

In the 1980s Néstor García Canclini asked, "Why are museums so bad in Latin America?" (116). His own response suggests that in one way or another, the problem was not simply economic or infrastructural but rather historical or political. The deficiency of Latin American museums, he argued, arose from the failure or incompleteness of national projects. He then singled out the Mexican case as the exception, the one country in the region capable of crafting a coherent national narrative after its early twentieth-century revolution and, as a result, capable of establishing high-quality museums. While this perspective closely aligns with Fredric Jameson's mandate from the same decade, which prescribed "national allegory" to all third-world cultural production—a concept we have since moved beyond—it is worth pondering whether the entire adventure of finding and rescuing *Metropolis* and then defending its authenticity could be explained by the nonexistence of a national cinematheque in Argentina.

I want to propose that this case sheds light on how the act of collecting is different when it is done in the global peripheries. I suggest that twenty-first-century Latin American documentaries are assembling a collection of collections that unfold a different agenda and reflect their dysfunctional contexts. While there may be infinite motivations for collecting, official collections such as state-sponsored museums, zoos, and libraries are founded on an attribution of value and importance that extends beyond the personal. These collections must reflect a public interest, and the preservation of their objects should be seen as a public good. In the absence of a systematic and rational policy for the development of such collections, or in the failure to do it efficiently, arises an individual, private impulse toward gathering objects that preserve stories and build a vicarious history. These personal repositories are only subjectively important and valuable and may lack actual material worth, as opposed to the objective and even monumental value held by official collections. The documentaries analyzed here engage with the act of collecting as both a cultural and a natural impulse, closely mirroring the practices of natural history, in which cataloging and preserving specimens served to organize and protect knowledge of the natural world. By drawing on this archival legacy, the films reveal how collecting extends beyond mere accumulation, becoming an act of memory and resistance within fragile contexts. This connection to natural history deepens the films' exploration of preservation, emphasizing how documenting endangered or ephemeral elements of nature and culture transforms the archival act into a statement on survival and continuity.

Locus of Collection

I will illustrate this contrast by comparing two works that represent similarly private, impulsive acts of collecting. Nonetheless, they stem from very different aspirations and material circumstances. I am taking the following description from the MoMA website of the video installation *Grosse Fatigue* (2013) by French visual artist Camille Henrot:

> She made the video while in residence at the Smithsonian Institution in Washington, D.C., digging into its vast collections to pull together images of objects and specimens, like animal skeletons and carved figurines, with footage she shot in offices and collection storage. Henrot merges this with additional video clips and images she both made and found online. Characterizing her structuring of *Grosse Fatigue* as "an experience of density itself," she frames this material in layered pop-up windows that continually open and close against the changing background of a computer desktop. Brief pauses in the pacing stand out in Henrot's otherwise rapid-fire sequencing. Sometimes, a woman's hands appear in the frames, nails playfully manicured to match the colorful backgrounds. A spoken word-style voiceover, which interweaves stories of creation from across cultures, structures the visual cacophony. Henrot describes this mash-up of scientific discovery and religious myth-

making as an "intuitive unfolding of knowledge," a presentation meant to highlight our abundance of information, as well as its limits. (MoMA)

Like many other works of contemporary art, *Grosse Fatigue* is challenging to classify in part because of the combination of footage shot by the artist for the film and images found elsewhere or its articulation as a work of "desktop cinema" but mainly because there is no clear protocol for its distribution and consumption.[2] It is usually presented in art shows and biennials as a video installation, but museums and collections also distribute it on their websites. Each experience is likely to be different and result in a distinct work. *Grosse Fatigue* delves into the complexities of documenting and categorizing extensive knowledge, utilizing natural specimens as its primary source material to create a layered, immersive audiovisual experience. This structure reflects the encyclopedic ambitions and inherent limitations of the documentaries discussed later. By juxtaposing chaotic visual and verbal narratives, Henrot's work raises questions about the construction, preservation, and interpretation of knowledge, concerns at the heart of these films' agendas.

The form of the work draws inspiration from the experience of using a computer operating system, which reserves its own space for each element, its own window, but also allows for the coexistence and overlapping of multiple windows, creating a controlled or preprogrammed chaos. This form is, in turn, modeled on the recursive property of language, the syntactical ability of structures to always contain a smaller structure within, ad infinitum. But the ambition of Henrot's thirteen-minute video is to "tell the story of the creation of the universe," according to the Guggenheim Museum presentation; it is an agenda most clearly articulated by the voiceover text but reinforced by many of the images (Guggenheim).

Even though the windows in *Grosse Fatigue* are parallel, meaning they are all contained in the same larger structure and not nested within one another, the work cleverly tricks viewers into thinking they are generated in a recursive pattern. In this way, Henrot aims to represent totality, inviting us to glimpse the absolute in the form of a computer screen, whether real or simulated. *Grosse Fatigue* can be seen as a contemporary revival of the eighteenth-century aspiration of encyclopedism. Her arbitrary collection of embalmed birds, books, bones, and random images suggests that everything can be known and collected.

João Moreira Salles's film *Santiago* (2007) centers around Santiago Badariotti Merlo, the butler who worked in the director's family home in Rio de Janeiro during the late twentieth century. This documentary narrates the story of the eccentric butler, the spectacular mid-century modern home in the affluent neighborhood of Gávea where he served, and the influential Moreira Salles family. It also delves into the director's numerous, albeit unsuccessful, attempts to tell this deeply personal story. Filmed during the twilight of Santiago's life, after he has retired to a small apartment, the family has dispersed, and the house is vacant, the film closely aligns with the reflections seen in the memory documentaries that became prevalent in the region at the turn of the century, dominating the first two decades of the twenty-first century. At one point in the film, Santiago explains to his neighbor what all the fuss created by the shooting crew in his small apartment is: they are "embalming" him, that is, the documentary, even though it is about a living man and his current life, carries the imprint of memory that is prevalent in Latin American documentary in the wake of dictatorship and other traumatic experiences. Even the present is treated as a matter of memory.

Santiago is a bizarre character portrayed in the film with a blend of curiosity and affection. Among his many eccentricities, such as dressing in a tuxedo to play Beethoven on the piano when alone, speaking a mix of Portuguese, Spanish, and Italian inadvertently, and performing a dance with his hands to classical music, it is Santiago's extensive compendium of what he calls his "mental abortions" that sheds light on the act of peripheral collecting. For decades, Santiago meticulously copied the history and genealogy of hundreds of royal families and aristocratic lineages from various origins, including western European, Middle Eastern, Indian, and Chinese, and spanning multiple historical periods, from ancient times to the twentieth century. He transcribed these texts using a typewriter on loose sheets, primarily standard copy paper and sometimes stationery from the hotels where he accompanied the family during their travels. These sheets often featured marginal handwritten notes, comments, highlights, and underlines. The pages were not bound but instead grouped by family or genealogy, tied together with ribbons, labeled, and neatly stacked in a bookcase. By the time the film was shot, Santiago had amassed tens of thousands of pages.

What is the point of this collection? Santiago's lifelong work undeniably stands as a triumph of will, a remarkably consistent effort to learn and commemorate, an act of love in itself. However, it serves no practical purpose; it is a capricious collection of quotes that lacks a method or system and is impractical for consultation, offering no novel information. In a sense, it is an encyclopedic endeavor, yet one that lacks the organization or coherence necessary for real utility. Santiago's collection is technically a book, but not one fit for publication. It is unlikely to be read and may not even be preserved. While it is a beautiful object, its value remains purely aesthetic.

Henrot and Moreira Salles exhibit two distinct inclinations when it comes to creating comprehensive collections. Henrot's work, with its presumption of universality, misleads observers to once again trust in the potential of an all-encompassing, absolute archive—essentially, the internet. Despite its title suggesting exhaustion and a sense of weariness, it remains a techno-optimistic work. However, the condition of possibility of *Grosse Fatigue* is not solely technological; it arises from Henrot's experience with the Smithsonian Institution, the world's largest physical repository of objects and information, encompassing an extensive collection of enormous museums. Henrot's audacity that she can narrate the story of the universe can only stem from this unique locus of enunciation.

In contrast, through the work of Santiago Badariotti, Moreira Salles documents a creative collection. It appears as a seemingly redundant and uncritical compilation of intensely personal and intimate preexisting materials and entirely disconnected from public interest. Nevertheless, it responds to an absence, an impulse to fill an empty and unfilled archive. Santiago assembled his collection as a traveler, a wandering individual. While some of the books he transcribed might have come from one of the twenty-one specialized libraries of the Smithsonian Institution, it is evident that his locus of enunciation is distinct from that of the Smithsonian.

Collections, Series, Hoards

Since the 2000s, Latin American documentary filmmakers have become curators of the most unusual collections perhaps because, in the end, "every documentarian is something of a collector" [todo documentalista tiene algo de coleccionista] (Speranza, "Colección"). In addition to Badariotti's "mental abortions," Patricio Guzmán's "subjective trilogy" (*Nostalgia de la luz,* 2010; *El botón de nácar,* 2015; and *La cordillera de los sueños,* 2019) presents an assortment of telescopes, observatories, stars, bones and human remains, seashells, photographs, words in Indigenous languages, rocks, and an impressive collection of VHS tapes that record popular protests during and after the Pinochet dictatorship. Walter Benjamin offers a useful characterization here: "What is decisive in collecting is that the object is detached from all its original functions in order to enter into the closest conceivable relation to things of the same kind. This relation is the diametric opposite of any utility, and falls into the peculiar category of completeness" (204). I want to highlight that these documentaries present the materials as either collections—that is, somehow organized, systematized, curated—or articulate them as collections. In the first case, the film presents the collector; in the second, the collectors are the filmmakers themselves. For example, while it would be a stretch to define a language as a collection of words, that is all we are given by Guzmán in *El botón de nácar* when each of the interviewed subjects is asked to translate a Spanish word chosen by the filmmaker into their Indigenous languages, Kawésqar or Alacaluf. Presented together in a sequence, these words become a collection.

The same is the case for Jonathan Perel's *Responsabilidad empresarial* (2020), a sixty-eight-minute film that consists solely of a series of static shots of factories, sugar refineries, and workshops that were complicit in the disappearance of their workers during the 1976–1983 dictatorship in Argentina. In a series of parallel shots at the Recoleta Cemetery in Buenos Aires, Nicolás Prividera's *Tierra de los padres* (2011) combines two collections: a series of tombs of prominent figures in Argentine history and readings of classic Argentine texts. Andrés di Tella's multiple collections throughout his films (photographs, fragments of Super-8 footage, quotations) peak with *327 cuadernos* (2015), a film about Ricardo Piglia's illness and death that heavily relies on the 327 notebooks of his personal diaries, a collection that titled the film. Finally, *Bonanza: En vías de extinción* (Ulises Rosell, 2001), described as "a merciful allegory of Argentina" [una alegoría piadosa de la Argentina] following the 2001 collapse (Speranza, "Colección"), is more overtly a film about a collector. Norberto Muchinsci lives in a scrapyard where he accumulates all sorts of discarded materials he deems worthy of reentering the marketplace. Everything has a price, including the animals he has managed to capture on the outskirts of Buenos Aires.

It would be excessive to say that these random, eccentric collections respond to the same anxiety imposed on Santiago. Still, it is hard to overlook the overt attention to the gathering of objects, their arrangement as organized series, and their oversaturation with symbolic and emotional meaning. I claim that these precarious, capricious, monstrous collections expose a defect or an excess of materiality and reveal patterns in historical trajectories that have been left out of official collections.

The Herbarium and Film

Leandro Listorti is a projectionist. He was for about a decade a programmer for the BAFICI (Buenos Aires Festival Internacional de Cine Independiente) and is the director of two of the most programmatic documentaries made from and on collections. His exceptional sensibility for the act of collecting may stem from his work as an archivist in the Museo del Cine in Buenos Aires and his affinity for the genre of "found footage," on which he has co-edited a book (Listorti and Trerotola). Listorti's first full-length "documentary," *La película infinita* (2018), is probably the film that best condenses the relations between archival obsession and cinephilia.

This fifty-eight-minute film was crafted by editing together outtakes, retakes, and screen tests from approximately fifteen Argentine films, footage that Listorti discovered within his workplace, the Museo del Cine. Some of these films were released, while others remained unrealized projects, including a few high-profile ones like Lucrecia Martel's *El eternauta* and Nicolás Sarquís's *Zama*. Some segments feature original sound, but the majority are silent. Additionally, there are some non-original sounds incorporated into the film. The film's defining characteristic lies in Listorti's approach to editing, purposefully spotlighting the diverse origins of the footage. He often emphasizes elements such as clapperboards, countdown sequences, and blank stock in film transitions. As is frequently observed in discussions about the possibility of non-narrative films, the mere sequential arrangement of these combined fragments introduces a temporal logic that may elicit decoding and interpretation akin to a narrative. However, the primary intent of the editing is to showcase its own process, ultimately presenting the film as a curated collection of fragments of films that never came to be. Thus, *La película infinita* does not represent or discuss a haphazard collection; it is, in essence, a haphazard collection. Moreover, as implied by the title's double meaning in Spanish as "infinite" and "unfinished," it is an open collection. The film's infinitude is derived from its capacity to encompass an endless array of random segments from a theoretically infinite number of films.

Listorti's keen interest in film preservation takes on a more reflective tone in *Herbaria* (2022). The title refers to the systematically arranged collections of dried plants that figure prominently in the film, old herbaria collected in Argentina in the late nineteenth century by German botanists and herbaria still under construction in different public institutions in the country. Most of the film shows images of plants in different stages of the collection process, when they are cut from the forest, placed between sheets of paper, pressed, arranged on the pages of the herbaria, used as a model for drawings, or scanned. Beyond the basic distinction pointed out by the US National Park Service when predicating these notions on nature, "conservation seeks the proper use of nature, while preservation seeks protection of nature from use." Listorti's initial reflection on the actions of conserving and preserving is more literal and has to do with plants and old herbaria. However, it immediately transitions to film, drawing parallels between the archive of dead plants and the archive of old films. *Herbaria* does not merely ask how these two acts of collection and preservation are alike or different. More interestingly, the film shows how they overlap.

Herbaria is a film about conservation. Its opening focus is on telling the story of the collection and preservation of Argentine flora in the dried form of herbaria. In a reasoning that seems to have arisen from the height of modernity, herbaria are heralded in the film as instruments for the conservation of plants, even if they have been killed in the act of collecting them. Then the film turns into a narrative on the preservation of historical Argentine herbaria, the first of them done by German and French naturalists who collected the samples and took them back to Europe for study and classification. Listorti quickly brings film and film preservation into the mix, telling this as a family story. Cristóbal Hicken (1875–1933), the first notable Argentine botanist, who collected more than 150,000 species, was the uncle of Pablo Ducrós Hicken (1903–1969), a notable collector of films and film equipment. The city of Buenos Aires commemorated both men in the names of the Escuela Municipal de Jardinería and the Museo del Cine, respectively. This deep interconnection takes the understated proposition that the medium of film can be an instrument for the conservation of herbaria and of plants, as exemplified by the film's last two shots. In the first, a preserved sepia sequence of a plant is shown with the superimposed title "This flower was filmed in 35 mm in 1912" [Esta flor fue filmada en 35 mm en el año 1912]. Then, a contemporary shot of a different flower has the title "This flower was filmed in 16 mm in 2021" [Esta flor fue filmada en 16 mm en el año 2021].[3]

The first text to appear on the screen reads, "Since 1750, around 500 species of plants have disappeared from the planet. This amount is more than double the sum of extinct birds, mammals, and amphibians" [Desde el año 1750 alrededor de 500 especies de plantas han desaparecido del planeta. Esta cantidad es más del doble que la suma de aves, mamíferos y anfibios extintos]. Toward the end, a new slide reads, "Since its birth at the end of the nineteenth century, it is estimated that between 80 to 90% of silent cinema has disappeared worldwide. And approximately only 50% of sound cinema in film survives to this day" [Desde su nacimiento a finales del siglo XIX, se estima que entre el 80 al 90% del cine mudo ha desaparecido alrededor del mundo. Y aproximadamente sólo un 50% del cine sonoro fílmico sobrevive hasta nuestros días]. In between, an unidentified voice, which could be attributed to an Argentine botanist or biologist, reflects,

> Within the discourse of conservation biology, (we ask) why conserve? The first answer is biodiversity: this set of species, genes, and ecosystems has intrinsic value. That is, it stands on its own. Value is always relational. There is a link between us and what we value . . . There are no easy answers to what to conserve or not. But the important thing is who decides, right? Who decides what is worth conserving?
>
> Dentro del discurso 'biología de la conservación, ¿por qué conservar?,' lo que aparece primero es la biodiversidad y es 'este conjunto de especies, genes, ecosistemas tiene valor intrínseco.' Es decir, vale por sí mis-

mo. Y el valor es siempre relacional. Existe un vínculo entre nosotros y aquello valorado. . . . No hay respuestas fáciles de qué conservar o no, pero la importancia de quién decide, ¿no? . . . quién decide qué vale la pena conservar y por qué.

These observations offer a mix of ideas around conservation. One is the notion that things (either species or films) will disappear if no action is taken. Another is the disparate warning about the extinction of species, which points to the conservation of nature, and the destruction of film, that is, the conservation of culture. Even if the film also states that "plants conserved from many years ago are useful, for example, to understand climate change in relation to human activity" [Las plantas conservadas desde hace años sirven, por ejemplo, para entender el cambio climático en relación a la actividad humana], there is an important distance between preventing a species from going extinct and preserving a hundred-year-old herbarium. Or a hundred-year-old silent film.

The question turns to technology as the film shows an array of ways in which herbaria are used today, from serving as models for scientific illustrations to being scanned to produce more valuable images closer to a live three-dimensional plant model. We are informed that the process of "identification, classification, and naming of specimens . . . has remained unchanged for the last 200 years" [identificación, clasificación y denominación de especímenes . . . se ha mantenido sin modificaciones durante los últimos 200 años]. The technique for gathering the specimens has not changed either: placing the sample between newspaper pages, pressing it, letting it dry, sewing it to a page for exhibition. This sequence registers another layer of conservation, the conservation of an art (Figure 12.1). There seems to be the suggestion that, when it comes to plants and herbaria, a peak has been reached and that the technical resources currently available are sufficient.

The situation is different when it comes to cinema. Film, an art form ushered in by a technological innovation, contains the seed of its own destruction. *Herbaria* comments extensively on the challenges of preserving film and on the intrinsic self-destructive nature of old nitrate stock, which is highly flammable and inextinguishable, as the chemical process of its combustion generates its own oxygen. Thus, conserving films is going against their nature; they are intrinsically transitory, and not even technology can save them, Listorti suggests. In what becomes the most beautiful metaphor of the inevitable course of time and deterioration in the film, he warns, "The following images no longer exist in physical media. The artisanal scanning process involved the total disintegration of the material" [Las siguientes imágenes ya no existen en soporte físico.

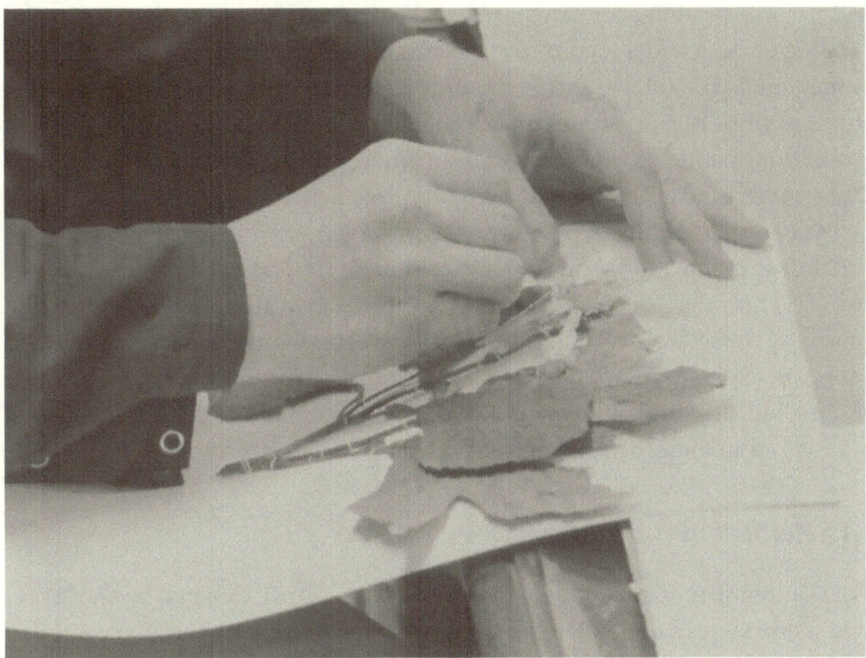

Figure 12.1. Leandro Listorti, still shot, *Herbaria,* 2022. Listorti documents the old art of making an herbarium. Courtesy of Leandro Listorti.

El proceso de escaneado artesanal implicó la desintegración total del material]. What appears next is a segment of a Basque pelota game in the south of Buenos Aires circa 1925 that produces a Phoenix-like effect (Figure 12.2).

The multilayered discourse on conservation in *Herbaria* seems to be dismissive of the classic opposition between nature and culture and invested in exploring possible collaborations or the blurring of the distinction. The film includes discussions on efforts to preserve biodiversity against the menace of climate change and the advance of monoculture, and thus it speaks to the meaning of the word "conservation." Argentine herbaria throughout history, the main focus of the film, are also a primary form of plant conservation; dead specimens, which are a valuable source of information available for consultation, can be utilized in many forms and may at some point be the only real evidence of the existence of a species if it goes extinct.

Film, on the other hand, is an invaluable tool for a different kind of plant conservation. As the final two shots of the film (the flowers dated 1912 and 2021) show, living plants can be preserved by registering them on film, turning them into "change mummified," as André Bazin termed it in the 1940s. The plants conserved in film can be vestiges of a real past in two ways: they are a record of a specific specimen that existed in the profilmic space in the past and no longer does and a record of a specific species that may not exist anymore sometime in the future, and whose only existence will be in the phantasmatic form of film. But the same can be said about herbaria, that recording these collections on film is a way of preserving them for the future, as these artifacts are also perishable and threatened by contingencies.

As Listorti reminds us, plants can be conserved as living species, in herbaria, in film, or in films about herbaria. Then there is the issue of film conservation, which is the other focus of Listorti here. Ultimately, it all comes to the question posed by the scientist: "Who gets to decide what is worth conserving?" Listorti does not explicitly ask or answer that question—regarding plants, herbaria, or films—but the whole film stands as a denunciation of the precarious state of conservation today of both culture and nature.

The Herbarium and the Forest

The Colombian documentary *Homo Botanicus* (2018), directed by Guillermo Quintero, stages a different relationship between cinema and herbaria, one that is very much about the present, even if anchored in a deep reflection on memory, the density of the past, and the passage of time. The film portrays Colombian botanist Julio Betancur, who has worked for decades as a researcher for the Colombian National Herbarium. By the time the film was made, Betancur had personally collected almost 20,000 species on trips to the different biomes in Colombia. The long first part of the documentary follows Betancur and the new apprentice Cristian Castro on an expedition to the Amazonian rainforest in southern Colombia. We see them in the task of gathering the samples in the forest, identifying them, laying them between sheets of newspaper, and storing them in their bags. It is an intense, solitary, and even dangerous endeavor. Still, the filmmaker makes sure to present it as an act of love and care, an attitude underlined when they speak of the plants they collect as their "amadas" (beloveds) or when, after being caught by surprise singing a love song, Castro explains that he is not in love with a woman but with the plants. The second, shorter part of the film shows them in Bogotá, working in the greenhouse where they keep some living specimens, and in the facilities of the Colombian National Herbarium, where the collected dried species end up.

The film speaks of Betancur and Castro's obsession with plants but also about the relationship between them, a mentorship analyzed from the perspective of someone who was in the apprentice's place before. Quintero introduces the film with a voiceover declaration of his past interest in botany and his relationship with Julio Betancur. Over images of the Colombian forests, Quintero says,

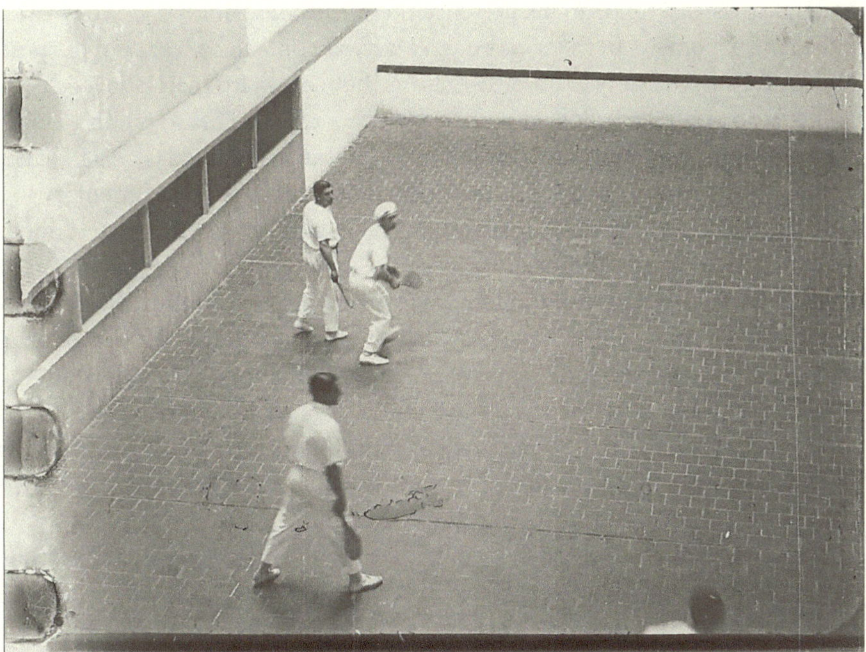

Figure 12.2. Historical shot of Basque pelota, Leandro Listorti, *Herbaria*, 2022. This footage from 1925 in Buenos Aires had to be destroyed for it to be preserved. Courtesy of Leandro Listorti.

Here I am again, at last. In the Andean cloud forests of Colombia. An image from my own past. An image from fifteen years ago. I can see myself again, surrounded by the tangle of greenery. A young student, forever following behind his professor, botanist Julio Betancur. And naturally, I think of the vertigo that forced me to leave it all behind. Faced with that infinite garden, why continue the endless task of classifying and collecting nature? And yet, now that I am here, I can't resist. I want to start all over again, call up Julio, and go away with him on a new expedition.

Finalmente, acá estoy, de vuelta, en estos bosques de niebla de los Andes colombianos. Una imagen de mi propio pasado. Una imagen de hace quince años. Puedo verme, otra vez, ahí metido entre esa maraña verde. Ese joven estudiante que seguía siempre a su profesor, el botánico Julio Betancur. Y claro, pienso también en ese vértigo que me hizo dejarlo todo. Ante este jardín infinito, ¿para qué seguir esa tarea eterna de colectar y clasificar a la naturaleza? Y sin embargo, ahora que estoy acá, no lo puedo resistir: querer comenzar de nuevo, llamar al profesor Julio y acompañarlo una vez más en una nueva expedición.

Soon afterward, to explain his presence there, Quintero observes, "They've allowed me to come with them, although I haven't come to collect as in the past, although I am here only to observe them as they search for their beloveds" [Me han dejado venir con ellos, aunque no venga a colectar como en un pasado. Aunque venga solo a observarlos mientras buscan a sus amadas]. The former botany student has become a filmmaker, and shooting this documentary is a new way to collect plants; it is the same trip, a similar expedition, a familiar location. As in the commonplace trope of the ethnographer entering the forest, time seems to go backward, and film is the instrument to capture it. The old apprentice failed and stopped believing in the importance of the task of collecting and classifying, the modern enterprise of seeing, grabbing, and understanding it all. He quit. But now, turned filmmaker, he comes back, not because he has recovered a lost faith but because he has found the beauty of the undertaking lived as a mission by Betancur and Castro. The filmmaker is not here to collect the plants. He will collect the collectors.

Betancur, Castro, and the film measure time through the act of collecting. Betancur announces that he will celebrate his 20,000th species with a party in the middle of the forest. We see a montage of old pictures of Betancur while the voiceover lists numbers and species: "8,975 *tillandsia biflora,* 18,555 *cyrtochilum betancurii,* 37 *heliconia samperiana.*" This is a device to illustrate the dreams and fantasies of his disciples, who imagine Betancur first naming all the plants he has collected, then all the species in the Amazon rainforest:

> Then, a bit drunk, we began imagining him again, in Santa María, Julio, the great scientist, reciting every species in the Amazon. And, of course, we imagined that as Julio recited every name, one by one, during that time several of the plants would have disappeared and new ones would be discovered. And by the time he reached the end, he'd have to start all over again.
>
> Y entonces ya algo ebrios, nos dio por imaginarlo de nuevo, ahí en Santa María. Julio, el gran hombre de ciencia citando esas plantas de toda la Amazonía. Y claro, nos dijimos que mientras Julio citara uno a uno esos nombres en todo ese tiempo, algunas de esas plantas ya habrían desaparecido, y otras se habrían descubierto. Y entonces al terminar, le tocaría volver a comenzar, y citar de nuevo.

Quintero had earlier warned of the endless task of classifying and collecting nature. The time and history outside the forest erupt from the ancillary materials, the old newspapers used to dry up the specimens. Newspapers are present in most of the shots, but only as paper; text is visible but never acknowledged until it becomes a figure for the passage of time. While placing leaves to dry, Castro says, reading the news on the page he is using, "Oh, I'd forgotten B. B. King is dead" [Ya no me acordaba que había fallecido B. B. King].

As the herbarium becomes an index of time for the men who built it and for the filmmaker who recorded the process, the power of this metaphor extends to the natural world itself: the herbarium is also a testimony of the time of the forest. As Castro states, "It's a bit nostalgic, isn't it? Let's take the census of this forest because who knows if it will still be here tomorrow" [Es un poco nostálgico, ¿no? Porque vamos a censar esto, este bosque, porque mañana no sabemos si esto va a estar acá]. Drawing back to the coupled concepts of conservation and preservation, Betancur's work—and the work of a long list of botanists named by the voiceover narrator, all the way back to Alexander von Humboldt—the herbarium is a repository of species that may not exist in the near future but also the mere witness to that existence in time: "It's impossible to know if a species has disappeared if it was never described" [Imposible saber si una especie desaparece sin antes haberla descrito]. Film here, in contrast to how it operates in *Herbaria,* is less preoccupied with the inventory and recording of the materiality of flora and herbaria. Instead, it is invested in capturing that nostalgia, the loving act of collecting all the specimens within an infinite forest. *Mutatis mutandis,* Listorti's old herbaria or Quiroga's herbaria in the works, which will eventually be the only trace left of the forest, recall what Jennifer Lynn Peterson has termed the Anthropocene viewing condition. "It is now common to experience an Anthropocene viewing condition," Peterson observes, "in which old images of nature cannot be separated from our new knowledge of the climate crisis . . . not only global warming but also mass extinction, habitat loss, and climate injustice" (20). Peterson claims that the climate crisis requires new ways of seeing and perceiving characterized by awareness of finitude. As revisiting old films of nature through the lens of climate change shapes a new structure of feeling around endangerment, the time-capsule nature of herbaria resembles the way film works with time and thus can trigger the same anxiety about environmental peril, loss, and extinction.

Herbaria in Listorti's film were also artifacts of nostalgia, but in that case, it is a nostalgia for the collection per se, immersed in a discourse of national history. In Quintero's *Homo Botanicus,* it is the forest that is subject to an ethos of nostalgia, a forest frozen in time by the herbaria that will be, in the future, their only representation. Nevertheless, there is a more radical way in which the conservation of the forest is performed through an act of collecting. Anaïs Taracena's 2017 documentary *Los médicos de la montaña* tells the story of the medical practices developed by rural and Indigenous peoples in the Guatemalan highlands during the civil war of the 1980s. Because of the de facto absence of the state in those regions, the network of *médicos de la montaña,* mountain physicians, has stayed in place until today and serves to showcase the value and effectiveness of community organizing. The film's first sequence introduces Shulin, an Indigenous practitioner of traditional medicine who is taking a stroll in the forest accompanied by a boy, probably an apprentice. Shulin is looking at the plants on his way, explaining, "In the mountain, we never know if we'll live or die, so we know a lot about plants to treat illnesses . . . a few wild plants that our ancestors taught us" [En la montaña como estamos en que vamos a vivir o vamos a morir, nosotros conocemos muchas plantas para curar las enfermedades . . . Algunas plantas silvestres que nuestros antepasados también nos han enseñado]. In their own way, these few wild plants taught by the ancestors form a collection. It is a virtual collection of living plants, specimens that need to exist in the forest and be available to those who can identify them and know how to put them to good use. There is a scale of interests, uses, and needs ranging from Listorti's preservation of a flower in film, to Betancur's preservation of samples of species that may go extinct, to Shulin's preservation of the knowledge about a plant that can stop an injured person's bleeding. But they all seem to articulate some sort of private, individual response to a collective need and a basic lack or deficiency that each film presents as a threat.

Shawn Miller has advocated for the rise of Latin American environmental awareness. This call encompasses more than a consciousness regarding environmental preservation; it also entails reexamining history to uncover the impulse *avant la lettre.* As Miller succinctly puts it, "Latin America has yet to unearth its environmental prophets—individuals disregarded by their contemporaries, consigned to obscurity without due recognition or commemoration" (203). Early in *Herbaria,* an unidentified voiceover tells the story of Argentine botanist Juan Aníbal Domínguez (1876–1946), who inherited from a French naturalist, the first one to work in Argentina, a collection of correspondence that included letters from Humboldt. In the early twentieth century, Domínguez traded this collection of letters to the Museum of Berlin for an herbaria collection created by nineteenth-century botanists who had been invited to Argentina by Domingo Faustino Sarmiento. These herbaria were duplicates of originals that stayed in Germany. During World War II, the originals were lost in the destruction of Berlin. Nowadays, the Argentine copies are valuable for being the first records of the local flora as well as the only samples left of the work of those prominent European botanists. This unexpected turn of events in which the copies take the place of the originals (as would happen later with Fritz Lang's film) and the very oppositions of original versus copy, center versus periphery are destabilized, speaks again of the asymmetry between Latin

America and metropolitan cultures. It also reveals once more that there is an emotional investment in the act of peripheral collecting.

Graciela Speranza's analysis of Gabriel Orozco's work *Asterisms* (2012) dialogues with this dynamic. Orozco's work comprises two parallel installations, *Sandstars* and *Astroturf Constellation*, each of them a collection of found objects. In the first, he gathered debris found on a beach in Baja California; in the second, small objects and trash items found on the artificial turf in a sports field in Pier 40, New York City. Speranza calls this a "cultural paradox":

> On the coasts of Mexico, his own space of belonging turned exotic in an uninhabited natural reserve, Orozco collected large-scale global debris that traveled thousands of kilometers pushed by the currents of the Pacific, while in a sports field near his home in New York, cosmopolitan center of the world par excellence, he gathered minuscule, hyperlocal remnants, synecdoches of American culture. The world, he seems to say in the play of scales, can be very small in a New York sports field and very large on a Mexican beach.

> En las costas de México, su propio espacio de pertenencia vuelto exótico en una reserva natural deshabitada, Orozco recolectó detritos globales de gran escala que viajaron miles de kilómetros empujados por las corrientes del Pacífico, mientras que en un campo de deportes próximo a su casa de Nueva York, centro cosmopolita del mundo por antonomasia, reunió restos ínfimos, hiperlocales, sinécdoques de la cultura norteamericana. El mundo, parece decir en el juego de escalas, puede ser muy pequeño en un campo de deportes neoyorquino y muy grande en una playa mexicana. (*Cronografías* 174)

The European herbaria and the old copy of *Metropolis* kept in Buenos Aires prove that Latin America's peripheral modernity, with its peripheral collections and collectors, is not premodeled by central economies and can counter and alter the extractive impulse imprinted in the origin of natural sciences, opting instead for becoming the locus of conservation.

From Herbarium to Sampler

But herbaria can be shown to have had a different function. Sofía Hansen's 2023 short film *Muestrario* tells the story of the devastation wrought by industrial extractivism in Hornopirén, a location and forest in southern Chile. About half of the twenty-six-minute film is devoted to images of botanical species collected and exhibited in old herbaria; the colors of the leaves have faded, and the paper on which they are displayed has yellowed. The sheets are labeled with institutions such as the United States National Museum, the United States National Herbarium, and the University of California. One is marked "Botanical Garden Expedition to the Andes, 1935–1936. Peru, Bolivia, Chile, Argentina. T. H. Goodspeed, Director." Entries and comments are typewritten in English; there are a few handwritten notes also in English. It is these comments and notes that reveal the actual purpose of the collections: "Valuable timber tree," "Economic uses," "Near small sawmill village," "Wood used for boats and oars."

Unidentified voices of interviewed villagers tell a fragmented story of the various forms of exploitation of the area's natural resources and their own adaptations to this economic history. Some moved there to work for a Japanese factory, others worked on fish farms. Some witnessed "intentional burning of forests in search of farmland for livestock and housing" [quemas intencionales de bosque por buscar suelo cultivable, para ganadería y para habitar]. Most shockingly, they refer to the place where they live as a "sacrifice zone" [una zona casi de sacrificio] where trees have been systematically felled over decades. "BIMA stands for Forestry and Timber Industries, and it was a national company that in the sixties joined forces with a company called Simpson Timber from the United States. Between the sixties and seventies was the peak of Alerce here in Chile" [BIMA se llama Bosque Industria Maderera y fue una empresa nacional que en su momento, en la época del sesenta, se juntó con una empresa que se llama Simpson and Turner, que era de Estados Unidos. Entre el sesenta y el setenta fue el apogeo máximo de alerce acá en Chile].

The title of the film reveals its distinct perspective: not herbarium but *muestrario,* sampler or collection of samples. It becomes clear that the scientific interest in the collected, classified, and named species (with comments on the translation of species names into English, revealing a preoccupation with accurately identifying resources as a necessary first step for efficient exploitation) is only secondary to the interest in their economic use. The beauty of the collection remains untainted, and the film seems to waver between aesthetic devotion to the herbaria and political denunciation of their purpose. At times, even the background sound reflects this tension, blurring the line between the soothing sound of water and the destructive sound of fire. The story told by *Muestrario* is painful and sad, yet the film lingers on breathtaking imagery of nature and landscapes along with old herbaria, maps, and living botanical species.

Shot in a style that recalls old home movies, through levels of exposure, treatment of color, and aspect ratio, the film is less programmatic than *Herbaria* and *Homo Botanicus* in establishing a direct correlation between building herbaria and making a film. The documentary opens, nonetheless, with a beautiful metaphor that strongly evokes the work of filmmaking:

> One would think that in the forest, which is very dense and has many trees, the light does not trespass, let's say, does not reach the ground, right? But incredibly, when the light begins to penetrate at different times of the day, new spaces are created, new images are created. You may think that the forests and the trees are fixed, and everything is fixed, but everything is in motion. It proves, let's say, the fact that it is alive.
>
> Uno podría pensar que el bosque, que es muy tupido y que tiene muchos árboles, la luz no traspasa, digamos, no llega al suelo, ¿no? Pero el bosque, increíblemente, cuando la luz empieza a penetrar a diferentes horas del día, se van creando nuevos espacios, se van creando nuevas imágenes. Uno puede pensar que los bosques están fijos y están los árboles ahí y todo es fijo, y está todo en movimiento. Como que se demuestra, digamos, el hecho de que está vivo.

Light and movement weave the magic of the forest. They are also the building blocks of cinema.

Notes

1 I would like to thank the participants of the Tulane Academic Writing Workshop for their insights and suggestions to improve this text.

2 Initially coined in the late 1990s as referring to movies "filmed, edited, and screened entirely with digital technologies" (Tryon 93), the term "desktop cinema" later evolved to designate "a movie that only takes place on the screens of the protagonist's digital devices" (Ugenti 178), or "films that incorporate the desktop environment in the narrative by way of a combination of pre-recorded desktop footage and other sources, including original or found footage, as well as PC-delivered data" (De Rosa and Strauven 249). As digitalization becomes the norm of archiving practices in the twenty-first century, it makes sense that collections can only be exhibited on a screen.

3 All English quotations are taken from the film's official subtitles and are not my own translations.

Works Cited

Barrera, Julieta. "La verdadera historia detrás del hallazgo de *Metrópolis* en Buenos Aires." *Infobae,* March 5, 2023. https://www.infobae.com/cultura/2023/03/05/la-verdadera-historia-del-hallazgo-de-metropolis-en-buenos-aires/.

Benjamin, Walter. *The Arcades Project.* Harvard University Press, 1999.

De Rosa, Miriam, and Wanda Strauven. "Screenic (Re) Orientations: Desktop, Tabletop, Tablet, Booklet, Touchscreen, etc." In *Screen Space Reconfigured,* edited by Susanne O. Saether and Synne T. Bull, 231–262. University of Amsterdam Press, 2020.

Di Tella, Andrés, dir. *327 cuadernos.* J. C. Fisner, 2016.

García Canclini, Néstor. *Hybrid Cultures: Strategies for Entering and Leaving Modernity.* Translated by Christopher L. Chiappari and Silvia L. Lopez. University of Minnesota Press, 1995.

Guggenheim. "Camille Henrot: *Grosse Fatigue.*" N.d. https://www.guggenheim.org/artwork/33581.

Hansen, Sofía, dir. *Muestrario.* Llilán, 2023.

Jameson, Fredric. "Third-World Literature in the Era of Multinational Capitalism." Social Text, no. 15 (1986): 65–88. https://doi.org/10.2307/466493.

Guzmán, Patricio, dir. *El botón de nácar.* Atacama Productions, 2015.

Guzmán, Patricio, dir. *La cordillera de los sueños.* Atacama Productions, 2019.

Guzmán, Patricio, dir. *Nostalgia de la luz.* Atacama Productions, 2010.

Listorti, Leandro, dir. *Herbaria.* Maravillacine, 2022.

Listorti, Leandro, dir. *La película infinita.* Maravillacine, 2018.

Listorti, Leandro, and Diego Trerotola, eds. *Cine encontrado ¿Qué es y adónde va el* found footage?. Buenos Aires Festival Internacional de Cine Independiente, 2010.

Miller, Shawn William. *An Environmental History of Latin America.* Cambridge University Press, 2007.

MoMA. "Camille Henrot: *Grosse Fatigue,* 2013." https://www.moma.org/collection/works/175938.

Peña, Fernando Martín. *Metrópolis.* Fan, 2011.

Perel, Jonathan, dir. *Responsabilidad empresarial.* BasaltFilm, 2020.

Peterson, Jennifer Lynn. "An Anthropocene Viewing Condition." *Representations* 157 (2022): 17–40.

Prividera, Nicolás, dir. *Tierra de los padres.* Trivial Media, 2014.

Quintero, Guillermo, dir. *Homo Botanicus.* Casatarántula, 2018.

Rosell, Ulises, dir. *Bonanza: En vías de extinción.* INCAA, 2009.

Speranza, Graciela. "Colección de colecciones," *Otra Parte,* 1 (2003). https://www.revistaotraparte.com/op/artes/coleccion-de-colecciones/.

Speranza, Graciela. *Cronografías: Arte y ficciones de un tiempo sin tiempo.* Anagrama, 2017.

Taracena, Anaïs, dir. *Médicos de la montaña.* Asombro, 2019.

Tryon, Chuck. *Reinventing Cinema: Movies in the Age of Media Convergence.* Rutgers University Press, 2009.

Ugenti, Elio. "Between Visibility and Media Performativity: The Role of Interface and Gesturality in Desktop Cinema." *Cinéma & Cie. Film and Media Studies Journal* 21, no. 36/37 (2021): 177–192.

13

Marias and Joões

Naming in Maria Esther Maciel's Poetic Encyclopedias

NATHANIEL WOLFSON

In this essay I set out to trace a path that eventually leads to Maria Esther Maciel's 2021 work *Pequena enciclopédia de seres comuns*, a poetic encyclopedia that shares certain attributes with premodern bestiaries. Across its entries and illustrations by artist Julia Panadés of various (mostly) real animals and plants—that is to say, beings that are already known to exist in the world—its descriptions read like lyrical narratives. Each entry is labeled by the official Latin binomial name as well as a common name, and entries are grouped according to the latter. The reader, moreover, encounters a nominal commons that sustains the book's organization and by extension, a set of relations established onomastically. Common rather than scientific names not only organize the encyclopedia's entries but constitute how species are said to exist in the world. We learn by reading and savoring Maciel's narrative entries that beings act like their names.

The names that populate her books are woven together in various intricate ways, repeating and echoing in the minds of the reader. Each book is carefully and lovingly sutured to the other. *Pequena enciclopédia,* for instance, begins with a dedication to the biologist and author Zenóbia, the invented protagonist of Maciel's earlier book *O livro de Zenóbia* (2004). I asked myself, was it in fact Zenóbia who authored the *Pequena enciclopédia?* Did Maciel's invented biologist, known for her fascination with lists and flora, sign her own book under the name "Maria Esther Maciel" in an act of forgery? These questions start to pile up as one reads her poems, beginning with *O livro de Zenóbia* and proceeding to the next, *O livro dos nomes* (2006).

In both those books, short narrative chapters of poetic prose give the reader a portrait of Zenóbia. Although written in the third person, the narrator of *O livro de Zenóbia* has intimate knowledge of Zenóbia's experiences of loss and desire, revealing the textures of a particular kind of mind experiencing the world. The book includes, toward the end, Zenóbia's own lists of plants and preferred recipes. Divided into numbered subsections, the chapters' numerical patterning suggests a rational design and a logic that recalls the planned architecture of João Cabral de Melo Neto's poetry, among others drawn to poetic systems. Although *O livro de Zenóbia* appears systematic, its prose constantly deviates from plot and argument. Dream sequences of interiors with piling detritus seem irreconcilable with the book's otherwise regular scaffolding.

Zenóbia returns in Maciel's next project, *O livro dos nomes,* as one entry among others in a compendium of names organized by alphabetical sequence. Each name—from A to Z—comes with an etymology as well as a prose narrative entry that dramatizes the named person. Regular alphabetical ordering, suggesting homogeneity and standardization, is disordered by the narrative's inner movements. Out of order, names appear and reappear across entries, allowing the reader to weave together affiliations, perceive connections, and sense a rhythm. This logic of filiation, independent of any preordained alphabetic sequence, pertains to what I am calling a nominal commons, a set of relations established according to common names.

Maciel is, in general, attuned to a range of poetic, political, and scholarly efforts to think alternatively about logic, rationality, and power by situating knowledge geographically, ecologically, and politically. In her scholarship on animality she addresses literary examples and visual art that reimagine the traditionally Western genres of encyclopedias, dictionaries, and bestiaries. For example, in her perceptive essay on Afro-Brazilian artist Arthur Bispo do Rosário she examines the encyclopedic lists and catalogs he sketched onto the sculptures he constructed during his internment in mental hospitals ("*A enciclopédia de Arthur Bispo do Rosário*"). This essay on Bispo do Rosário sheds light on how her own poetry configures the organization of the natural world, offering alternative and nonrational methods of naming and classifying. Several of her characters are experts in homeopathy and healing and maintain deep ancestral skills. A great aunt who Zenóbia adopted as a godmother possesses "an almost shy elegance" [uma elegância quase tímida] and a wisdom not verifiable or fact-based: "each of her gestures contained a logic, a wisdom, a measure" [cada um de seus gestos continha uma lógica, um siso, uma medida] (*O livro de Zenóbia* 31).[1]

In what follows, I explore a path of Maciel's writing by attending to three of her books of poems. This path will end up on her book most related to natural history, scientific nomenclatures, and the conventions of bestiaries. If this aspect of her work is most interesting to the reader, I invite them to begin with this essay's final section and work backward. They'll face no risk of losing the trail, since Maciel's work itself has evolved nonlinearly. Each of her books is entwined into her others, producing in turn a rich matrix of names, characters, and stories. It is my preference to begin with *Zenóbia* in order to set up my commentary on the anti-extractivist poetic operations exemplified in Maciel's more recent encyclopedia. One might be tempted to separate her more ecological poems that feature nonhuman species from those that cultivate human and human-nonhuman relationships. I resist this temptation, tracing instead the desire of the name as it weaves through common spaces produced by a language that is not entirely human. Along the way, I acknowledge a few debates on the theory of names. Maciel's poetic practice of naming—with proper names, common names, and scientific names—does something highly imaginative with nomenclatures and classifications. Her poems reveal how naming bears on ecology and natural history, discourses that have, over the past few decades, become increasingly connected to poststructuralist theories of language.[2] But rather than dismiss language as such, as some recent materialist philosophies of nature polemically suggest is necessary (Barad 132), Maciel insists on the ongoing importance of words even in projects that radically critique humanism or negate "the human" in the name of a broader, less hierarchical, and otherwise radically more dynamic conception of life.

The Unclassifiable

Maciel's scholarship investigates fictive inventories of nature, texts that engage human and nonhuman relations, all subsumed within a genre that she explores called "zoopoetics."[3] She has chronicled these tendencies within art and literature, displaying her devotion to a wide range of aesthetic objects from various language and cultural traditions. Unlike many of these creative practices, her own fiction does not feature fantastical or invented beings. Instead, she revels in the delightful surprises that exist within the realm of existing and polyonymous plants and animals. In this sense, her artistic work stands out from most others.' Her scholarship, instead, addresses the fantastical dimensions of zoopoetics, including its creative borrowings from natural history writing and colonial documents.

In her article "Imagens zoológicas da América Latina" (2001), Maciel describes works of fiction that draw on documents produced by the first colonizers of the land that became known as South America, produced before the triumph of scientific rationalism in the eighteenth century, when Swedish biologist Carl

Linnaeus created a binomial system of taxonomy. These early modern texts belong to a time when colonizers who set out to name and collect specimens from the New World lacked the nomenclature to scientifically acknowledge the radical otherness of the animals and plants they observed. Zoological writings, therefore, drew upon fabulous content from medieval bestiaries and the teachings of ancient precursors of zoology, such as Aristotle's *History of Animals* and Pliny the Elder's encyclopedic books of *Natural History*.

Among examples of early twentieth-century texts that profile colonial-era representations of Brazil's fauna are the writings of Afonso d'Escragnolle-Taunay, especially his *Zoologia fantástica do Brasil* (1900) and *Monstros e monstrengos do Brasil* (1937). His sources included classics such as the anonymous *Diálogos das grandezas do Brasil* (1618) as well as reports by sailors such as the Portuguese Pero Vaz de Caminha and the German Hans Staden. Contemporary authors in conversation with premodern models imagine hybrid, unclassifiable, and monstrous facts of nature.[4] In *Jardim Zoológico* (1999), the Brazilian poet and novelist Wilson Bueno creates a bestiary involving invented animals that, according to Maciel, display an animal reason and knowledge deeply rooted in the body. Bueno's invented "giromas," for instance, have dozens of eyes that listen, breathe, and excrete liquid, thereby fertilizing themselves (Maciel, "Imagens" 95). The "rememorantes" are a species of night elves possessed with incredible memory and nourished by human dreams. Bueno's bestiary revisits and inverts processes of colonial domestication, imagining in the process, Maciel proposes, "another model of Latin American reason" [um outro modelo de razão latino-americana] (Maciel, "Imagens" 93). She explains further that this "savage reason" [razão selvagem] points, at the same time, to a kind of logic that "destabilizes the legitimized models of rationality that an imposed Westernization forced us to adopt as our own" [desestabiliza os modelos legitimados de racionalidade que a ocidentalização imposta nos obrigou a adotar como nossos] (93).

In another essay, "Nas fronteiras do humano e do não humano: Poéticas da natureza na literatura brasileira do século XXI," Maciel focuses on contemporary literary practices that critique destructive forms of extractivism and accelerating ecological destruction. Thinkers Ailton Krenak, Eduardo Viveiros de Castro, Dominique Lestel, and Donna Haraway inspire the fiction of contemporary Brazilian authors like Nicanor Sena, Astrid Cabral, Olga Savary, Sérgio Medeiros, and Josely Vianna Baptista. These writers invoke Indigenous epistemologies as well as diverse literary references from multiple geographies and periods. Maciel shares many interests with these contemporary references, though her characteristic focus on naming sets her apart. In what follows, I explore the function of naming in two of her early books of poems. The reader invested in her later more explicitly ecocritical poems or those closer to the encyclopedic or bestiary genres may choose to skip over this section or, alternatively, to linger with it.

Stolen Words

O livro de Zenóbia is dedicated to Zenóbia, a character who reemerges *en abyme* throughout Maciel's oeuvre. A dizzying likeness exists between the author, narrator, and character. We learn from the narrator that the eponymous character has a penchant for forgery. She also evidently shares several interests with the author, who, in turn, reproduces her character's diary: lists of weeds, "perplexing fish," "rare cities," spices and fragrant herbs, endangered birds, orchids and bromeliads, favorite words, and more. These inventories gesture to the contents of Maciel's fictive encyclopedias as well as to the falsifying acts of authors like J. M. Coetzee and Fernando Pessoa, whose pseudonyms and heteronyms similarly act as subterfuges for seduction and deception.

Maciel discloses the fabricability of names when she writes, "Alicia Mirabilis was the name Zenóbia used when she published her book on the miracles of Saint Clare. When writing about Teresa of Avila, she chose the name Notylia. She discovered it while researching orchids on an island she invented on the day of her greatest joy" [Alicia Mirabilis foi o nome que Zenóbia usou quando publicou o seu livro sobre os milagres de Santa Clara. Já ao escrever sobre Teresa d'Avila, escolheu o nome Notylia. Descobriu-o quando pesquisava orquídeas numa ilha que inventou no dia de sua maior alegria] (*O livro de Zenóbia* 35). One can easily appreciate why Zenóbia chooses to rename herself Alicia Mirabilis, after a species of sea anemone that changes shape as night falls, expanding its column and tentacles to catch its food. Nature, neither pure nor innocent, shares with poetry the magic and utility of semblance.[5] The ruse of Zenóbia's discovery of the orchid genus is an open secret among others. Words are often revealed to be things that can be falsified, gifted, or even stolen. "In 1987, Zenóbia met a woman in the street who sold words. They were all made up. She fell in love with 'ilágrime.' She discreetly bought it. However, she later learned that it was a stolen word" [Em 1987, Zenóbia conheceu na rua uma mulher que vendia palabras. Eram todas inventadas. Encantou-se com "ilágrime." Comprou-a, sem alarde. Porém, mais tarde, soube que essa era uma palavra roubada] (35). Zenóbia, like other characters, discreetly steals words as if they are prized possessions. The neologism "ilágrime" is particularly valuable, the passage suggests, because it is ever so slightly fake.

Numbers and naming work together to summon a chimerical order of things. In the same chapter, the narrator describes Zenóbia as so grief-stricken upon the death of her first dog that "she wrote his name on the floor a hundred times six times" (escreveu cem vezes seis vezes o nome dele no chão) (33). Two unequal numbers listed side by side convey the existence of a parallel reality. The broader context of the phrase concerns human and nonhuman animal love and with it, the pathos of creaturely death, a topic explored in the chapter "Rastros e pêlos" (Traces and Hairs). The unlikely timing of death and birth, of aging and rejuvenation, present alternatives to the telos of development and destruction. The action of writing a name is imbued with intense pain as well as desire; Zenóbia collects names, as do other characters. One man, described in a chapter of short fiction likely authored by Zenóbia herself, constantly peruses phonebooks to collect the names of women that begin with the letter A. "One day, for no visible or laughable reason, he gave up the search. Nights and nights of insomnia followed: that's when he discovered, in despair, that the biggest challenge was to find a woman with no name, born—preferably—in the month of February" [Um dia, sem qualquer motivo visível ou risível, desistiu da pesquisa. Noites e noites de insónia se sucederam: foi quando descobriu, em desespero, que o maior desafio era achar uma mulher sem nome, nascida—de preferência—no mês de fevereiro] (125). This apparent non sequitur, like others in the book, derails narrative flow and concentration, seeming to point nowhere except to an anonymous male's desire to find, possess, and ascribe to others the perfect name. In Maciel's next project, *O livro dos nomes*, Zenóbia dreams that her mother "asked her for a new name, on the grounds that the one she had was false and too dense" [lhe pedia um novo nome, sob a alegação de que o que tinha era falso e denso em demasia] (168).

O livro de Zenóbia is constructed as a series of chapters, most of which are subdivided into four or six numbered parts. The first chapter portrays her at various ages: eighteen, thirty-two, forty, fifty-eight, seventy-four, and eighty-two. Chronological progression suggests development and progress only to be undone by the following chapter's discussions of the protagonist's various deaths or "false beginnings" [falsos começos]. "And then she died for the first time. Three more deaths followed in the space of sixty-eight days. Until her father found a book that taught him not to let her die anymore. Of the deaths she suffered, Zenóbia only keeps the memory of the first one: and she would write about it all her life" [E aí ela morreu pela primeira vez. Três outras mortes lhe sucederam desde então, em um espaço de sessenta e oito dias. Até que seu pai encontrou um livro que o ensinou a não mais deixa-lá morrer. Das mortes que teve, Zenóbia guarda apenas a memória da primeira: e sobre esta não deixaria de escrever a vida inteira] (7). Zenóbia's father here, as well as other fathers in Maciel's next book, offer different kinds of care and attention. However, they tend not to pass on their names to their daughters.

Names appear to have a strangely mimetic quality. Sometimes they resemble those whom they designate. Other times they impact how lives are lived. Early on in the book the narrator states, regarding Zenóbia, that "her life was the summary of her name" [sua vida era o resumo de seu nome] (15). In the chapter dedicated to her family, the reader learns of her brother, deceased before birth, whose "name never existed, although there was talk of Felipe or Matheus" [nome nunca existiu, embora falassem em Felipe ou Matheus] (23). On the other hand, the name of her sister Dolores invokes the pain of her birth. Pain, in general, "exists in the exact measure of our name, although it always has a blank element, which never comes back to memory or dreams" [fica na exata medida de nosso nome, embora tenha sempre um elemento em branco, que nunca volta à memória ou aos sonhos] (19). If pain is both fleeting and equivalent to the dimensions of a name, the latter seems to persist beyond lived experience. Names, in other words, seem fixed and permanent, while experiences, especially those related to the body, are quite the opposite.

But this is only partially true. In fact, in other parts of the book, we discover an entirely different account of names that, like experience, shift according to a person's fleeting sensuous perceptions. Names, in some situations, might measure lives, but in other contexts, they are not required for living. Some people have no names to speak of. In the chapter "Family Album" [Álbum de família], Zenóbia's grandmothers and adopted elders remain nameless. Maciel here invokes heredity while questioning the idea that a name is passed down patronymically across generations. From one grandmother Zenóbia receives "her greatest inheritance" [sua maior herança]: recipes for regional sweets [quitandas] (27). From her adopted grandmother—among other alternative filiations ascribed to her—Zenóbia acquires words that "were always luminous, made of a rare knowledge, the kind that seems invented" [palavras eram sempre claridades. Feitas de um saber raro, desses que mais parecem inventados] (29). Among these words, the most precious collectively build up to a phrase: "Only those who have been loved by a dog know what a gift is" [Só quem já foi amado por um cão sabe o que é uma dávida] (29).

Words, like the relationships between humans and dogs, are gifts. In a chapter of short narrative passages seemingly unconnected to the plot of the novel, we encounter a woman who abandons her family name as well as her earthly possessions. Accordingly, names traffic as objects that circulate like things. With this character Maciel seems to imagine what the anthropologist William Mazzarella describes as "the ideal of a name resistant to all trafficking, to all truck and barter." In his article "On the Im/Propriety of Brand Names," Mazzarella addresses philosophies of naming including a minor but significant passage from *Minima moralia* (1951), in which Theodor Adorno "observes that a *truly* proper name would be self-identical to the point of utter non-identity. It would be *useless* in the best sense—that is to say, it would be entirely unavailable for any human purpose." In what follows I continue to explore a few theories of the proper names that obliquely touch upon Maciel's poetic practice.

Maciel doesn't attempt to build up to a grand philosophy of language with the tools of her poetry. The "pequena" (small) in the title of her *Pequena enciclopédia de seres comuns* (Small Encyclopedia of Common Beings) suggests an ironic desire to abbreviate and delimit something impossibly capacious. That being said, readers of her poetry might wonder how her encyclopedic entries refer to the people and animals they index. The reader might ask how the proper name summarizes or abbreviates a life. These questions bear on debates on naming, a few of which I proceed to address.

Theories of the Name

Among the most cited theorists of names is Saul Kripke, who in *Naming and Necessity* (1980) proposes the idea of the so-called rigid designator. Kripke argues that proper names, as well as other kinds of names, are "fixed" to specific references, thereby guaranteeing the identity between word and object in all possible worlds and all counterfactual situations. He writes, "Let's call something a rigid designator if in every possible world it designates the same object, a nonrigid or accidental designator if that is not the case" (Kripke 48). With this, he departs from other "descriptive" theories of proper names, for which names elicit descriptions or sets of descriptions that the name satisfies. By rejecting a descriptive approach, Kripke's rigid designator gives proper names an almost religiously upheld social power: names are created by what he calls an "initial baptism."

For several reasons critics have repeatedly questioned Kripke's baptismal scene of communication and understanding. Excess of various social and semantic causes often threatens to upset the self-identity of the proper name, to destabilize its reference. Kripke's rigid designation implies a language user who has been correctly initiated into the use of a name, passed down generationally, through a pact that secures its appropriate fixing of a referent to a name. Judith Butler takes up this point in her discussion of Kripke in *Bodies That Matter* (1993), in which she discusses the "illusory permanence" of the patronymic name and a related kind of violence in naming, the violence of interpellation and subjectivation. Maciel's poetic understanding of names can fruitfully converse with Butler's *Bodies That Matter,* especially where Butler discusses the literary "appropriation and displacement of the patronym" in Willa Cather's fiction. Cather "displaces the social basis of its identity-conferring function and leaves the question of the referent open as a site of contested gendered and sexual meaning" (Butler 154). We can recall here the named lineages and nonpatriarchal cycles of life in *O livro de Zenóbia*. Maciel suggestively uses first or given rather than second or inherited names to refer to her characters throughout her work.

Mazzarella cites examples of anthropologists of religion and magic who have troubled the notion of rigid designation in sacred contexts. These include Emile Durkheim's argument about totemism in his seminal *The Elementary Forms of Religious Life* from 1912, that the power of the totemic sign does not depend on it referring to the animals or plants from which it takes its name or to the human individuals who identify with a particular totem (Mazzarella). We might ourselves draw on the more contemporary writings of Brazilian anthropologist Eduardo Viveiros de Castro, who has written on the resistance to proper names in Amerindian cosmological thought and practice. In his essay "Cosmological Deixis and Amerindian Perspectivism," Viveiros de Castro discusses the widespread avoidance of self-reference on the level of personal onomastics: names are not spoken by bearers nor in their presence; to name is to externalize, to separate (from) the subject. Proper names, like those referring to ethnic groups, are imposed from the outside. He stipulates that "the majority of Amerindian ethnonyms which enter the literature are not self-designations, but rather names (frequently pejorative) conferred by other groups: ethnonymic objectivation is primordially applied to others. . . . Ethnonyms are names of third parties; they belong to the category of 'they' not to the category of 'we'" (Viveiros de Castro 476). Viveiros de Castro, in broad strokes, discusses the perspectival pragmatics of Amerindian cosmology whereby names refer to the condition of personhood that can include animals and spirits. The tremendous power of naming attributes to human and nonhuman beings the capacities of conscious intentionality and agency. Without relinquishing proper names altogether, Maciel's poetry involves strategies that trouble the hierarchies made possible by rigid designation and other facets of language philosophy that strictly delimit how and what can be named.

Z for Zenóbia

Following the release of *O livro de Zenóbia*, Maciel went further in exploring the vicissitudes of poetic naming. Zenóbia returns in *O livro dos nomes* as one entry among others in a compendium of names organized by alphabetical sequence. Each name comes with an etymology as well as a prose narrative portrait that dramatizes the named person's life. Across its entries, names emerge and return outside of the book's ostensive ordering. The reader is not asked to go from beginning to end to encounter a linear story. Rather, the book's design tests the reader's capacity to remember and to draw connections between disconnected accounts of life events. The order of reading, therefore, is unfixed. The reader aggregates and disaggregates information all the time, retracing steps in order to find the threads that tie names together and potentially weave stories about family histories, friendships, lovers, and various kinds of human and nonhuman relationships.

O livro dos nomes abides by an approximately similar stylistic structure as Maciel's later *Pequena enciclopédia.* Each entry includes a title composed of a proper name as well as a short poetic subtitle. These are followed by an etymology, fact, or definition in the form of a statement of public knowledge such as "it is known that" [é sabido que], derived from dictionaries of names and other sources, some real and others fictional.[6] Then a narrative, divided into four numbered parts, addresses the life of the person named. These entries, in turn, cite other proper names, some of which are cataloged in the book's entries. The outcome is an open structure, the kind that appears to grow and infinitely reproduce itself despite the existence of a finite alphabet. This feature of continuous growth corresponds with the description of Zenóbia in Z. "She pays attention to everything she breathes and learned from a friend that fate likes repetitions, variants, and symmetries. She is attracted to things that are unfinished, because she believes that not everything that begins ends" [Presta atenção em tudo o que respira, e aprendeu com um amigo que o destino gosta de repetições, variantes e simetrias. Tem uma atração pelas coisas inconclusas, por achar que nem tudo o que começa termina] (164–165).

By citing portraiture as Zenóbia's preferred genre Maciel recalls the presence of visual representations in encyclopedias and bestiaries. We should also remember the author's frequent collaborations with visual artists and illustrators. In Z, we learn a lot more about Zenóbia. We discover, for example, that she is the daughter of a doctor named Aristides and an elementary school teacher named Firmina. She prefers acidic colors such as yellow, red, and orange. She possesses many notebooks in which she writes lists of names, sketches of short stories, descriptions of dreams and people, recipes, and zoological curiosities. Her notebooks also contain information on the meanings of names from several kinds of sources, including onomastic dictionaries and old encyclopedias. She consults books of hagiography, such as Jacobus de Voragine's *Golden Legend* from the thirteenth century, as well as the *Etymologiae* of Saint Isidore of Seville (gifted to her by her friend Pliny, whose name recalls Pliny the Elder). Though she draws on legitimate sources, sometimes offering precise references, she invents the origins of the rarest names.

There is another crucial detail in Z worth noting: Zenóbia intended to dedicate herself to writing a book of names and to finish it by mid-2007. As she imagined her future project, she thought about emulating other writers of verbal portraiture, authors like Schwob, Cioran, Machado de Assis, Flaubert, and Borges. However, when she went to retrieve her notebook from a cupboard to begin writing, she discovered that it had disappeared. Desperately looking for it in every corner, she had to conclude that it had been stolen. *O livro dos nomes* is evidently, potentially, the lost manuscript. Several characters are possible suspects of theft and plagiarism.

The Common and the Proper

Pequena enciclopédia de seres comuns begins with an ontological premise that puts the status of the book in crisis.

> This book may not exist. Or rather, its nonexistence is what probably justifies it as a book. It was written by a biologist who is not a biologist but pretends to be one, in the way of the impossible. On the other hand, the living beings included in it—all classified according to certain peculiarities of their common names—have an irrefutable reality: whether by science, literature, or neither.
>
> Este livro talvez não exista. Ou melhor: sua inexistência é o que, provavelmente, o justifica enquanto livro. Foi escrito por uma bióloga que não é bióloga, mas finge ser uma, na medida do impossível. Já os seres vivos nele incluídos—todos classificados segundo certas peculiaridades de seus nomes comuns—têm uma realidade irrefutável: seja pela ciência, pela literatura ou por nenhuma das duas. (*Pequena enciclopédia* 5)

An unstable beginning—an assertion that is also an anti-assertion—sets the stage for a bestiary on species whose common names *appear* as proper names. Maciel's poetic experiment on being and naming asks whether one can extract from encyclopedias something much more speculative and fundamentally less restrictive and truth-oriented than registering nature by fixing it with the tools of language. Many of her entries begin with an apophantic "is" statement defining the being—"It is a fish" [É um peixe]—only to present a list of alternative names of the same species followed by a prose narrative that undoes certainty whatsoever.

Maciel appropriates natural history texts, botanical and zoological documents, and literary fiction to play with encyclopedias' taxonomic and semiotic conditions. The book's playfully encyclopedic nature aligns with other examples of contemporary conceptual poetry invoking onomastics.[7] In *Pequena enciclopédia*, names produce a nominal commons that houses an assortment of species—plants, nonhuman animals, and human beings. These species fall into sections entitled "Marias," "Johns" [Joões], "Widows and Widowers" [Viuvas e viuvinhas], and "Hybrids" [Híbridos]. Each entry includes a tag with a capitalized common name accompanied by a scientific or Latin cognate placed in smaller font, italics, and parentheses below, and each entry is accompanied by an illustration by Julia Panadés.

To address the book's title, it may be useful to recall the definition of common and proper names. Generally speaking, a common name or noun designates any being or thing within a class of beings or things. These pertain to a group and may, in English, occur with limiting modifiers such as the articles *a* or *an*, *some*, *every*, and *my*, or demonstrative pronouns *this*, *that*, and *these*. Proper names or nouns, on the other hand, are the designations of a particular person, place, or object. Though Maciel tags each entry with a common name, each of these reads as exquisitely unique and appears as if it were proper. Maciel leaves these names entirely capitalized, making any determination of name category based on typographic setting impossible. The ambivalence of category she establishes bears on the semantic question of survival. In several of the entries, the speaker accounts for the endangerment of the species. Catastrophe, in a sense, would literally mean reducing the common name to the proper name by limiting the species to an indivisible one. Common names appear like proper names in part because their commonness is constantly and increasingly imperiled.

The undecidable status of the name plays out in descriptive entries that further articulate the untruthful language we have noted throughout Maciel's oeuvre. Consider the entry "Maria-Cavaleira," the bird whose scientific name is *Myiarchus ferox*. In addition to describing in detail the color and shading of her beak, tail, and backside, the sounds she omits, and other features of her behavior, the entry includes inferences and invented facts such as that she sings alto only on Fridays and that only children can understand her song. Maciel mixes factual information from discussions of anatomy and feeding behaviors with content borrowed from popular knowledge or taken from literary sources. Though descriptive, the phrasing of her entries is measured and rhythmically consistent. At times, however, rhythm and tone promptly change when an "I" erupts from the description. The speaker's uses of deictic markers like "essa" (that)—rather than "esta" (this) or "aquela" (that)—variably measure the distance between the name and the referent, almost physically marking the intimacy between the being and the entry that defines it.

By way of a conclusion, I will briefly examine two further examples. In an entry on the fish species called "Maria-Luisa" (*Paralongchurus brasiliensis*), Maciel draws on the poetic possibilities of catachresis. The name Maria-Luísa appears to properly refer to a human being, but the entry's description creates possible confusion between fish and human animal. This nominal and descriptive fungibility not only reinforces the evolutionary relations between fish and humans but also highlights the shared environments in which both beings (of that same name) live: the polluted waters of "zona litorânea" (coastal zone). Recalling João Cabral de Melo Neto's *O cão sem plumas* (1950), the entry asks whether the river or the name binds these human and fish species together.

Lastly, consider the final entry in the section of "Marias" on the imagined human species "Maria-vai-com-outras." This is one of only two entries concerned with human animals. The name "Maria-vai-com-outras," however, is not invented; it is an expression commonly used to pejoratively describe someone whose actions are overly motivated by others.[8] Maciel recalls this expression with the line "She's a supportive human being. If she goes with others, it's above all to help them" [É uma humana solidária. Se ela vai com as outras, e sobretudo para ajudá-las] (*Pequena enciclopédia* 32). This unassuming but powerful critique of neoliberal values of self-sufficiency and individuality plays out onomastically. Upon rereading that line, it becomes evident that the name for humanity not confined or delimited by the ideals of an autonomous and proper self converges on its own description. In the second sentence, "Se ela vai com as outras," name and description quite literally converge and unfold. In a flash, the name unifies with its entirely proper self, becoming a strangely hybrid species within the space of a living paragraph. As a textual unit, the name has a certain plasticity or material quality, producing semantic thickness. It is as if the naming process of the entire book were here condensed into a single potent name: "Maria-vai-com-outras."

At first glance, Maciel's parody of encyclopedias and bestiaries may appear to reinstate the violence of systems in which names act like tools of control and restriction. However, in her poetic worlds, naming engenders several different kinds of undetermined mutual relations between species. Her poems deny the isolation of nature from society; for this reason, she is not, in my view, a pastoral poet. That is not to say that she professes a world of ecological stability. Maciel's projects all acknowledge, in some form, a state of precarity and possible ecological collapse. In her *Pequena enciclopédia de seres comuns*, the reader encounters direct references to pollution and resource depletion, as if names were needed as mnemonics for possible extinction. But this kind of direct commentary on crisis is not dominant in her work. The interplay

Maciel explores between definition and narrative exhibits an anti-extractivist style that sustains her nominal commons. Names are not imposed on subjects but maintain a sacred power. Lyricism prevails against banality, engendering subtle feelings, deeply sensitive reactions, and pleasure activated in daily tasks like gardening and cooking. A radical openness emerges here through tactile vocabularies, images of impossible spaces, and incomplete definitions that push concepts and words to the limits of their semantic meaning.

Notes

1 Translations of Portuguese in this chapter are by the author.
2 Among other sources, see Verena Andermatt Conley's *Ecopolitics* and Sue Ellen Campbell's "Land and Language of Desire."
3 In *Literatura e animalidade* (13–19), Maciel develops the notion of "zooliteratura" and "zoopoética."
4 Recent scholarship on aesthetics and ecology includes Joanna Page's *Decolonial Ecologies* (2023) and the present volume.
5 On the tropes of natural purity that sustain the false environmentalism of polluting corporations, see François.
6 Among Maciel's sources is Rosario Farani Mansur Guérios's *Dicionário etimológico de nomes e sobrenomes* (1949).
7 For a study of the conceptual genre of dictionary poetics, see Dworkin.
8 The other example of an invented human animal is the entry on the "Viuvinha-humana" (*Homo sapiens viuvensis*) discussed by Eduardo Jorge de Oliveira in his review of Maciel's book, "Uma enciclopédia comum e impossível."

Works Cited

Andermatt Conley, Verena. *Ecopolitics: The Environment in Poststructuralist Thought.* Routledge, 1997.

Barad, Karen. *Meeting the Universe Halfway: Quantum Physics and the Entanglement of Matter and Meaning.* Duke University Press, 2007.

Bueno, Wilson. *Jardim Zoológico.* Iluminuras, 1999.

Butler, Judith. *Bodies That Matter: On the Discursive Limits of Sex.* Routledge, 1990.

Campbell, Sue Ellen. "The Land and Language of Desire: Where Deep Ecology and Post-Structuralism Meet." *Western American Literature* 24, no. 3 (1989): 199–211.

de Oliveira, Eduardo Jorge. "Uma enciclopédia comum e impossível." *Revista Brasileira de Literatura Comparada* 23, no. 44 (2021): 270–273.

Dworkin, Craig. *Dictionary Poetics: Towards a Radical Lexography.* Fordham University Press, 2020.

François, Anne-Lise. "'O Happy Living Things': Frankenfoods and the Bounds of Wordsworthian Natural Piety." *diacritics* 33, no. 2 (2005): 42–70.

Kripke, Saul. *Naming and Necessity.* Harvard University Press, 1980.

Maciel, Maria Esther. "*A enciclopédia de Arthur Bispo do Rosário.*" *Outra travessia,* no. 7 (2008): 117–124.

Maciel, Maria Esther. "Imagens zoológicas da América Latina." *Niterói,* no. 10 (2001): 89–96.

Maciel, Maria Esther. *Literatura e animalidade.* Civilização Brasileira, 2016.

Maciel, Maria Esther. "Nas fronteiras do humano e do não humano: Poéticas da natureza na literatura brasileira do século XXI." *Aisthesis,* no. 70 (2021): 531–542.

Maciel, Maria Esther. *O livro de Zenóbia.* Lamparina, 2004.

Maciel, Maria Esther. *O livro dos nomes.* Companhia das Letras, 2006.

Maciel, Maria Esther. *Pequena enciclopédia de seres comuns.* Todavia, 2021.

Mazzarella, William. "On the Im/Propriety of Brand Names." *South Asia Multidisciplinary Academic Journal* 12 (2015). https://journals.openedition.org/samaj/3986.

Page, Joanna. *Decolonial Ecologies: The Reinvention of Natural History in Latin American Art.* Open Book, 2023.

Viveiros de Castro, Eduardo. "Cosmological Deixis and Amerindian Perspectivism." *Journal of the Royal Anthropological Institute* 4, no. 3 (September 1998): 469–448.

Afterword

Joanna Page

Latin America has emerged in recent years as one of the most important centers in the world for new environmental thought. It is a thought that is inseparable from activism, wrought as it has been in the intense struggles over land rights and ecological devastation that have spanned the continent. Such conflicts are not merely about access to natural resources but about the imposition of "ecologically inappropriate and culturally alien" models that strip habitats of their biodiversity and threaten the very way of life of local communities (Leff, "Power-Knowledge" 244). The environmental, social, and cultural destruction that has been carried out on a barely imaginable scale since the early colonial period has made Latin America, as the editors of this volume state, "a strategic place to think about the asymmetrical impacts of the climate crisis" (Nicolás Campisi and Lucas Mertehikian, this volume). This thinking has taken the form of a powerful critique of neoliberal capitalism as well as the promotion of alternative socioecological philosophies and practices, partly inspired by Indigenous principles of sustainability and reciprocity within more-than-human communities.

Within this matrix, many recent writers, artists, curators, and researchers have critically reexamined the discourses and collecting practices of natural history as a European discipline that came of age in the many botanical and zoological studies carried out in eighteenth- and nineteenth-century expeditions to Latin America. This has led to a rich corpus of recent works that unpick the imperial power relations inherent in natural history but also, at the same time, find in its practices something of real value for our present era of environmental crisis. Some of the essays in this volume focus in part on questions of representation: how artworks or exhibitions may, in drawing on the practices of natural history, either avoid or reinforce colonial stereotypes and hierarchies. But a much greater concern emerges with the environmental politics of the forms and institutions of art and natural history themselves.

Despite their claims to reflect and promote sustainable and decolonial practices, our museums, galleries, and other institutions, dependent as they frequently are on corporate funding, are often disappointingly embedded in the same economic and social model they wish to challenge (Valeria Meiller). The precarious state of many museums and archives in Latin America continues to reflect their peripheral position in a global symbolic and economic order (Antonio Gómez). Their very existence constitutes an affront to the natural habitats that they replace with concrete; their locations may encourage unsustainable forms of transport or reproduce spatial inequalities, as the Papalote Children's Museum in Mexico City does (Emily Hind). Part of the critique Hind develops here is directed at the absurdity and hypocrisy of the museum's call for children to solve an ecological crisis by playing with plastic toy bricks; this forms part of a broader argument that the museum as an institution can offer no "benign exterior" to increasingly urgent environmental crises.

In many ways, then, we can understand the recent boom in natural history in Latin American culture as a new chapter in the evolution of institutional critique. But the intense scrutiny to which we subject our institutions and spaces of culture has, at its heart, a belief that they are not only repositories of artifacts and bastions of privilege. Museums, collections, and the texts in which these are inscribed may also help to reactivate ancestral knowledge around biodiversity in the Global South (Luciana Martins), allow us to grasp the possibility of other futures, beyond global capitalism in its current expression (Florencia Garramuño), articulate the value of local modes of resistance by marginalized groups (Meiller), obey a logic of "peripheral collecting" that expresses different values to those of the center (Gómez) or give expression to the sacred and the lyrical in ways that run counter to the strict nomenclature of encyclopedias (Nathaniel Wolfson). They are places where old and new stories can be told, both of which are vital to rethinking and reimagining pasts and futures. These are especially important at a time when more of us are starting to believe that accounting for and addressing the environmental crisis is not a task that can be left to scientists alone (Meiller) and that natural history itself cannot answer the question "*Who* decides what is worth conserving?" (Gómez, my emphasis).

Ecocriticism, new materialism, and posthumanism have encouraged us to understand the natural world not as a mere backdrop or stage on which human dramas are played out, but as the material expression, over millennia, of interactions between different human and nonhuman forces. How to read these "land archives" (Carlos Fonseca), this "narrated matter" (Gisela Heffes), becomes one of the most intriguing questions for both literature and criticism in the twenty-first century. The revelation that nature is "nothing but the concretization of time" (Fonseca) or that "everything is made of time" (Gabriel

Giorgi) underlies the recurrent concern with temporality which shapes many of the works analyzed in this volume, as well as the approaches of their critics. A geological approach to the flows, accumulations, and eruptions of history allows us to understand how bodies, other types of matter, culture, and power have interacted in space over time. It also points to the possibility, as Gabriel Giorgi argues, of a conception of time structured by the stratum, oriented around "blocks of time that accumulate, overlap, and become contiguous without the promise of unification or totalization," bringing nonhuman temporalities into focus alongside human history. This thinking fractures the time of modernity and the nation-state "as a monopoly of collective temporalities" (Giorgi).

The task that many contemporary writers and artists set for themselves is to learn to read in the sedimentations of a landscape the changes—either slow or sudden—that shape interaction between human and nonhuman forces over timescales we struggle to conceive. For Matylda Figlerowicz, the challenge is how to rework literary form in such a way as to "make space for the nonhuman world," which involves a shaking-up of the hierarchies and divisions that structure modernity (and museums). Such concerns are not necessarily new, as Florencia Malbrán reminds us in a reading of Emilio Renart here, and Gisela Heffes also does in her return to Horacio Quiroga. Reaching back into the past to uncover texts that now seem prescient in their treatment of ecological damage or posthumanist sensibilities is an act that reminds us of something we must never forget. While discourses of the Anthropocene are relatively new, the environmental catastrophe they recognize is not. When we describe extinctions in the future tense, we run the risk of erasing the many extinctions already suffered by Black and Indigenous peoples in the past (Yusoff 51).

A critical engagement with natural history on the part of many recent artists, curators, and researchers effectively challenges the West's continued investment in a particular kind of knowledge, knowledge-by-accumulation, which is at the heart of modern European science (Marcaida López 37). The visual politics of baroque natural history collections—in which the feathers, fossils, shells, and bones of many different species shared a space of seeming equality with works of human artistry—are founded on death; only in death, and only through the exercise of power and capital, could such a diverse range of species be brought together in the same space. The cabinet of curiosities (reading on from Jerónimo Duarte-Riascos's reflections in this volume) thus flaunts not only human technical virtuosity but its capacity for destruction.

Against the catastrophism of much ecocritical writing, however, the essays in this volume mark a turn toward imagining futures that, if not exactly utopian, open up new horizons of possibility. This constitutes an important shift, as Fonseca describes it, away from approaching nature ontologically, as "the space of origins, authenticity, and being," and toward an understanding of nature as a "contested space where political narratives are articulated and disarticulated." This turn to politics is crucial, as it opens up the possibility of difference and change, of "other ways of inhabiting the planet" (Giorgi). As Enrique Leff argues, "Politics is the means by which . . . it is possible to move from a global world, governed by the unifying power of the market, to constructing a diverse world—different modes of being-in-the-world and of inhabiting the planet—steered by an ontology of diversity, difference and otherness" ("Political Ecology" 231). It is the basis on which we might start to consider the rights of all beings to construct "life-worlds" and to approach the question of what should be conserved and who has the right to make such decisions.

Acknowledging that our institutions and their practices are decisive political actors—and not merely spaces or forms of representation—is a crucial starting point, as is recognizing that they are irretrievably entangled with the forces they contest. For this reason, discourses of utopianism and ideological purity have no purchase here. Learning to live better with the other species that share this planet is a messy exercise, full of compromises. There is no universal model, no moral high ground, no simple cause and effect. Making life better for some communities or species usually means reducing the options for others.

That is why the emphasis on possible futures is so welcome here: it addresses the urgent task, in the context of global environmental catastrophe, of working out different ways of living on this planet. Art and literature have as much a role to play here as scientific predictions. If nature is really frozen time, then art and literature are also forms of sedimentation that contain within them not just the secrets of the past but the seeds of the future, just as seed banks preserve ancient DNA for times to come or the air bubbles trapped in Antarctic ice foretold an age that we could not yet see. What can we find, what can we recover and preserve, that might help us make sense of what is taking place, or even guide us toward a different path? It is in this spirit that the writers, artists, curators, and researchers represented in this volume have returned to natural history as a source of contested practices that throw new light on the increasingly contested spaces in which we live.

Works Cited

Leff, Enrique. "Political Ecology: A Latin American Perspective." *Desenvolvimento e Meio Ambiente* 35, no. 1 (January 2015): 29–64.

Leff, Enrique. "Power-Knowledge Relations in the Field of Political Ecology." *Ambiente e Sociedade* 20, no. 3 (July–September 2017): 225–256.

Marcaida López, José Ramón. *Arte y ciencia en el barroco español: Historia natural, coleccionismo y cultural visual.* Fundación Focus-Abengoa and Marcial Pons, 2014.

Yusoff, Kathryn. *A Billion Black Anthropocenes or None.* University of Minnesota Press, 2018.

CONTRIBUTORS

Nicolás Campisi is assistant professor in the Department of Spanish and Portuguese at Georgetown University. He is the author of *The Return of the Contemporary: The Latin American Novel in the End Times.*

Jerónimo Duarte-Riascos is assistant professor of Latin American and Iberian cultures at Columbia University in New York City and has previously taught at Northwestern University and the University of Chicago. He is a founding member of the curatorial collective de cabeza.

Matylda Figlerowicz is assistant professor of comparative literature at Harvard University. She is the author of *La memoria en construcción* (2015). Her most recent book project studies multilingualism as a form of critical theory.

Carlos Fonseca is the author of three novels, *Coronel Lágrimas, Museo animal,* and *Austral,* all published in Spanish by Anagrama and in English by Restless Books and Farrar, Straus, and Giroux. His academic monograph *The Literature of Catastrophe: Nature, Disaster, and Revolution in Latin America* appeared in Bloomsbury in 2020.

Florencia Garramuño teaches at the Universidad de San Andrés in Buenos Aires. Her published works include the books *Genealogías culturales: Argentina, Brasil y Uruguay en la novela contemporánea, 1980–1990* (1997), *Modernidades primitivas: Tango, samba y nación* (2007), *La experiencia opaca: Literatura y desencanto* (2009), *Mundos en común: Ensayos sobre la inespecifidad en el arte* (2015), *Brasil caníbal: Entre la bossa nova y la extrema derecha* (2019), and *La vida impropia: Anonimato y singularidad* (2022).

Gabriel Giorgi is professor at New York University. His published works include the books *Sueños de exterminio: Homosexualidad y representación en la literatura argentina contemporánea* (2004), *Formas comunes: Animalidad, biopolítica, cultura* (2014; translated into Portuguese in 2016), and in collaboration with Ana Kiffer, *Las vueltas del odio: Gestos, escrituras, políticas* (2020; first published in Brazil in 2019). He also coedited with Fermin Rodríguez the anthology *Excesos de vida: Ensayos sobre biopolítica* (2007).

Antonio Gómez teaches courses on Latin American culture, literature, and cinema at Tulane University and is associate editor of *Latin American Research Review.* He is the author of *Escribir el espacio ausente* (2014) and the coeditor of *The Film Archipelago: Islands in Latin American Film* (2022).

Gisela Heffes is a writer and professor of Latin American literature and culture at Johns Hopkins University. Her most recent publications include *Visualizing Loss in Latin America: Biopolitics, Waste, and the Urban Environment* (2023), and the coedited volumes *Turbar la quietud* (2023), *Un gabinete para el futuro* (2022), *Pushing Past the Human in Latin American Cinema* (2021), and *The Latin American Ecocultural Reader* (2020). She is also the author of the novel *Cocodrilos en la noche* (2020, 2023) and the collection of texts *Aquí no hubo ni una estrella* (2023), as well as the poetry book *El cero móvil de su boca / The Mobile Zero of Its Mouth* (2020). Her work has been translated into Swedish, German, French, Portuguese, and English.

Emily Hind is a professor of Spanish at the University of Florida. She is the author of *Dude Lit: Mexican Men Writing and Performing Competence, 1955–2012* (2019) and the book of interviews *Literatura infantil y juvenil: Entrevistas* (2020), which gathers twenty-two conversations with writers and editors of children's and young adult literature in Mexico.

Florencia Malbrán is a curator and writer. She has held curatorial positions at the Museum of Latin American Art, MALBA; the Guggenheim Museums in New York, Bilbao, and Venice; the Museum of Modern Art in Buenos Aires; and the Pinacoteca do Estado in São Paulo. She teaches at NYU's Gallatin School of Individualized Study in New York City and the Universidad de San Andrés in Argentina, where she is associate professor. Her book, *La prueba del presente: Ensayos sobre arte contemporáneo,* was published in 2023.

Luciana Martins is professor of Latin American visual cultures at Birkbeck, University of London, and a visiting researcher at the Royal Botanic Gardens, Kew. Her published works include the books *O Rio de Janeiro dos viajantes: O olhar britânico* (2001); *Tropical Visions in an Age of Empire* (with F. Driver, 2005); and *Photography and Documentary Film in the Making of Modern Brazil* (2013). Her forthcoming monograph entitled *Drawing Together: The Visual Archive of Expeditionary Travel* is supported by the Leverhulme Trust.

Valeria Meiller is assistant professor in the Department of Hispanic Languages and Literatures and a core faculty member of the Native and Indigenous Studies Initiative at Stony Brook University. She is working toward two book projects, *Necroterritories: Slaughterhouses and the Politics of Death* and *In Defense of the*

Land: Biodiversity and Linguistic Diversity in 21st-Century Poetry from Abiayala. Her work has appeared in journals such as *ISLE* and *Revista Hispánica Moderna.* She is also the director of Ruge el Bosque, an environmental project that has edited two plurilingual poetry anthologies dedicated to the Southern Cone and Mesoamerica.

Lucas Mertehikian is director of the Humanities Institute at the New York Botanical Garden.

Joanna Page is professor of Latin American studies at the University of Cambridge. She is the author of six books, including *Decolonial Ecologies: The Reinvention of Natural History in Latin American Art* (2023). She is also the coeditor of *Geopolitics, Culture, and the Scientific Imaginary in Latin America* (with María Blanco, 2020).

Ignacio Pastén López is a doctoral candidate in the Latin American, Iberian, and Latin Cultures Program (LAILaC) at the Graduate Center of the City University Of New York. He is an adjunct lecturer at the City College of New York and the College of Staten Island.

Ignacio Veraguas Caripan is a doctoral student in Spanish and Portuguese at Johns Hopkins University. He is the coauthor, with Pablo Chiuminatto, of the essay "Gusto, sabor y saber" (2021).

Nathaniel Wolfson is associate professor of Spanish and Portuguese and affiliated faculty of the Program in Critical Theory and the Berkeley Center for New Media at the University of California, Berkeley. He is the author of *Concrete Encoded: Poetry, Design, and the Cybernetic Imaginary in Brazil* (2025).

INDEX

Page numbers in *italics* indicate illustrations.

www.ingramcontent.com/pod-product-compliance
Lightning Source LLC
LaVergne TN
LVHW050955080826
845145LV00006B/1502

* 9 7 8 1 6 8 3 4 0 5 6 9 6 *